More Praise for *Against the Grain:*

"I admire *Against the Grain* very much. It is the most confirming and clarifying book about agriculture that I have read in a long time."—WENDELL BERRY, as written in a letter to James C. Scott

"This book is fascinating and original, containing a lesson on every page. Brilliant. James Scott is a legend."—TIM HARFORD, author of *Messy* and *The Undercover Economist*

"Fascinating. . . . Our agrarian-biased view of history, Scott concludes, could use some reworking. Most of the world's early human populations likely enjoyed semisettled, semiagrarian lives beyond the state's grasp."—SUZANNE SHABLOVSKY, *Science*

"Scott offers an alternative to the conventional narrative that is altogether more fascinating, not least in the way it omits any self-congratulation about human achievement. His account of the deep past doesn't purport to be definitive, but it is surely more accurate than the one we are used to."—STEVEN MITHEN, *London Review of Books*

"Fascinating."—GEORGE MONBIOT, *The Guardian*

"[*Against the Grain*] presents a comprehensive and convincing case that the transition from hunter-gatherer nomadism to permanent, agriculturally dependent settlements was a complete disaster for humankind. . . . Whatever your political leanings, the implications of Scott's book are as fascinating as they are wide-ranging."—WILL COLLINS, *American Conservative*

"The most interesting non-fiction read of the year. . . . Urgently recommended, and fun to read as well."—TYLER COWEN, *Marginal Revolution*

"*Against the Grain* delivers not only a darker story but also a broad understanding of the forces that shaped the formation of states and why they collapsed—right up to the industrial age. . . . An excellent book."—BEN COLLYER, *New Scientist*

"Fascinating. . . . Thinkers like Scott remind us that who we thought we are might not necessarily be the case. Such knowledge is empowering."—DEREK BERES, *Big Think*

"For more than 40 years, James Scott has written about those who resist being incorporated into political-economic systems. . . . In a provocative new book, *Against the Grain*, Scott now challenges us to rethink legends about the state and its origins."—JACOB LEVY, *Reason*

"Forget the Paleo Diet: Scott goes all the way in showing how early nomadic peoples in the Fertile Crescent were fitter, happier and more productive than the semi-enslaved ziggurat-builders of the ancient Mesopotamian cities."—JAMES WHIPPLE (M.E.S.H.), *Frieze*

"The story of a complex pattern of grain agriculture, early states, forced labor, and the extraction of surplus, and how all of these things were connected in ways that researchers previously never suspected. . . . Scott is a writer of extraordinary talent. . . . The constant interplay between the present and the distant past is one of the most appealing aspects of this book."
—JASON KUZNICKI, *Cato Journal*

"*Against the Grain* delivers what it says on the tin and is a fine piece of historical counter-narrative, with elements of environmental history woven throughout. . . . This results in a book that is fascinating, readable, but above all thought-provoking. It certainly made me ponder the 'civil' part in civilization."—LEON VLIEGER, Natural History Book Service

"A contemporary master of the political counter-narrative has produced a book on the origins of civilization—this is, quite simply, a must-read."—DAVID WENGROW, author of *What Makes Civilization?*

"This is a brilliant, accessible, and highly original account of the origins of sedentism, farming, states, and the relations between agrarian and nomadic communities. It should attract a wider audience than any of Scott's earlier books."—J. R. MCNEILL, co-author of *The Great Acceleration: An Environmental History of the Anthropocene since 1945*

"A sweeping and provocative look at the 'rise of civilization,' focusing particularly on those parts, peoples, and issues that are normally overlooked in conventional historical narratives."
—ALISON BETTS, University of Sydney

"Brilliant, sparkling, dissident scholarship. In Scott's hands, agriculture looks like a terrible choice, states and empires look fragile, ephemeral, and predatory, and the 'barbarians' beyond their borders lived in relative freedom and affluence."—DAVID CHRISTIAN, Macquarie University, Sydney

"Once again James Scott upends conventional wisdom with this riveting account of the entangled relations of humans, plants, and animals that created the first agrarian states. A provocative and eye-opening analysis of the dark side of our civilized world."—PETER C. PERDUE, author of *China Marches West: The Qing Conquest of Central Eurasia*

"This is a telling and timely argument at a time when the Anthropocene is beset by the harrowing histories of migrants and refugees who claim the rights of mobility in a desperate search for a life of security that is as urgent as it is elusive."—HOMI BHABHA, Harvard University

"*Against the Grain* is not just a "counternarrative," an outsider's skeptical reaction to received wisdom about the evolution of agricultural systems and the first states in Mesopotamia. Vainglorious kings with their generals and armies, sycophantic scribes, and royal architects and engineers are not Scott's heroes. His concerns are with urban laborers, peasants, and barbarians and the cleavage planes of resistance to rulers. Those studying Mesopotamia—and other early states—take heed."—NORMAN YOFFEE, editor of *Early Cities in Comparative Perspective*

Against the Grain

Against the Grain

A Deep History of
the Earliest States

JAMES C. SCOTT

Yale
UNIVERSITY PRESS
New Haven and London

Published with assistance from the foundation established in memory of Amasa Stone Mather of the Class of 1907, Yale College.

Yale University Press books may be purchased in quantity for educational, business, or promotional use. For information, please e-mail sales.press@yale.edu (U.S. office) or sales@yaleup.co.uk (U.K. office).

Set in Janson and Monotype Van Dijck types by Tseng Information Systems, Inc.
Printed in the United States of America.

Library of Congress Control Number: 2016960155
ISBN 978-0-300-18291-0 (hardcover : alk. paper)
ISBN 978-0-300-24021-4 (pbk.)

A catalogue record for this book is available from the British Library.

10 9 8 7 6 5 4

For my grandkids headed deeper into the Anthropocene

Lillian Louise
Graeme Orwell
Anya Juliet
Ezra David
Winifred Daisy

Claude Lévi-Strauss wrote thus:

Writing appears to be necessary for the centralized, stratified state to reproduce itself. . . . Writing is a strange thing. . . . The one phenomenon which has invariably accompanied it is the formation of cities and empires: the integration into a political system, that is to say, of a considerable number of individuals . . . into a hierarchy of castes and classes. . . . It seems to favor rather the exploitation than the enlightenment of mankind.

Contents

Preface

What you will find here is a trespasser's reconnaissance report. Let me explain. I was asked to give two Tanner Lectures at Harvard in 2011. The request was flattering, but having just finished an arduous book, I was enjoying a welcome spell of "free reading" with no particular aim in mind. What could I possibly do in four months that might be interesting? Casting about for a manageable theme, I thought about the two opening lectures I had been in the habit of giving to a graduate course on agrarian societies for the past two decades. They covered the history of domestications and the agrarian structure of the earliest states. Although they had gradually evolved, I was aware that they were woefully out of date. Perhaps, I reasoned, I could hurl myself at the more recent work on domestication and the earliest states and at least write two lectures that would reflect newer scholarship and be more worthy of my discerning students.

Was I ever in for a surprise! The preparation for the

lecture upset a great deal of what I thought I knew and exposed me to a host of new debates and findings that I realized I would have to put under my belt to do justice to the topic. The actual lectures, therefore, served more to register my astonishment at the amount of received wisdom that had to be thoroughly reexamined than to attempt that reexamination itself. Homi Bhabha, my host, selected three astute commentators—Arthur Kleinman, Partha Chatterjee, and Veena Das—who, in a seminar following the lectures, convinced me that my arguments were not remotely ready for prime time. Only five years later did I emerge with a draft that I thought was well founded and provocative.

This book thus reflects my effort to dig deeper. It is still very much the work of an amateur. Though I am a card-carrying political scientist and an anthropologist and environmentalist by courtesy, this endeavor has required working at the junction of prehistory, archaeology, ancient history, and anthropology. Not having any particular expertise in any of these fields, I can justly be accused of hubris. My excuse—which may not amount to a justification—for trespassing is threefold. First, there is the advantage of the naïveté I bring to the enterprise! Unlike a specialist immersed in the closely argued debates in their fields, I began with most of the same unexamined assumptions about the domestication of plants and animals, of sedentism, of early population centers, and of the first states that those of us who have not been paying much attention to new knowledge of the past two decades or so are apt to have taken for granted. In this respect, my ignorance and subsequent wide-eyed surprise at how much of what I thought I knew was wrong might be an advantage in writing for an audience that starts out with the same misconceptions. Second, I have made a conscientious effort, as a consumer, to understand the recent knowledge and debates in biology, epidemiology, archaeology, ancient history, demography, and environmental history that bear on these issues. And finally, I bring a background of two decades trying to understand the logic of modern state power (*Seeing Like a State*) as well as the practices of nonstate peoples, especially in Southeast Asia, who have, until recently, evaded absorption by states (*The Art of Not Being Governed*).

This is, therefore, a self-consciously derivative project. It creates no new knowledge of its own but aims, at its most ambitious, to "connect the dots" of existing knowledge in ways that may be illuminating or suggestive. The astonishing advances in our understanding over the past decades have served to radically revise or totally reverse what we thought we knew about the first "civilizations" in the Mesopotamian alluvium and elsewhere. We thought (most of us anyway) that the domestication of plants and animals led directly to sedentism and fixed-field agriculture. It turns out that sedentism long preceded evidence of plant and animal domestication and that both sedentism and domestication were in place at least four millennia before anything like agricultural villages appeared. Sedentism and the first appearance of towns were typically seen to be the effect of irrigation and of states. It turns out that both are, instead, usually the product of wetland abundance. We thought that sedentism and cultivation led directly to state formation, yet states pop up only long after fixed-field agriculture appears. Agriculture, it was assumed, was a great step forward in human well-being, nutrition, and leisure. Something like the opposite was initially the case. The state and early civilizations were often seen as attractive magnets, drawing people in by virtue of their luxury, culture, and opportunities. In fact, the early states had to capture and hold much of their population by forms of bondage and were plagued by the epidemics of crowding. The early states were fragile and liable to collapse, but the ensuing "dark ages" may often have marked an actual improvement in human welfare. Finally, there is a strong case to be made that life outside the state—life as a "barbarian"—may often have been materially easier, freer, and healthier than life at least for nonelites inside civilization.

I am under no illusion that what I have written here will be the last word on domestication, on early state formation, or on the relation between early states and the people of their hinterlands. My goal is twofold: first, the more modest one of condensing the best knowledge we have of these matters and then suggesting what it implies for state formation and for both the human and ecological consequences of the state form. By itself, this is a tall order and I have tried to emulate the standard set for this genre by the likes of Charles Mann (*1491*) and Elizabeth Kolbert (*The Sixth Extinction*). My second aim, for which my native trackers should be held blameless, is to draw larger and more suggestive implications that I imagine would be "good to think with." Thus I suggest that the broadest understanding of domestication as control over reproduction might be applied not only to fire, plants, and animals but also to slaves, state subjects, and women in the patriarchal family. I propose that the cereal grains have unique characteristics such that they would be, virtually everywhere, the major tax commodity essential to early state building. I believe that we may have grossly underestimated the importance of the (infectious) diseases of crowding in the demographic fragility of the early state. Unlike many historians, I wonder whether the frequent abandonment of early state centers might often have been a boon to the health and safety of their populations rather than a "dark age" signaling the collapse of a civilization. And finally, I ask whether those populations that remained outside state centers for millennia after the first states were established may not have remained there (or fled there) because they found conditions better. All of these implications I draw from my reading of the evidence are meant to be provocations. They are intended to stimulate further reflection and research. Where I have been stumped, I try to indicate so frankly. Where the evidence is thin and I stray into speculation, I try to signal that as well.

A word about geography and historical periods is in order. My focus is almost entirely on Mesopotamia, and in particular the "southern alluvium" south of contemporary Basra. The reason for this focus is that this area between the Tigris and Euphrates (Sumer) was *the heartland of the first "pristine" states in the world*—though it was not the location of the first sedentism, the first evidence of domesticated crops, or even the first proto-urban towns. The historical period I cover (aside from the deep history of domestication) encompasses the Ubaid Period, beginning roughly in 6,500 BCE, through the Old

Babylonian Period, ending roughly in 1,600 BCE. The conventional subdivisions (some earlier dates are disputed) are:

Ubaid (6,500–3,800 BCE)
Uruk (4,000–3,100)
Jemdet Nasr (3,100–2,900)
Early Dynastic (2,900–2,335)
Akkadian (2,334–2,193)
Ur III (2,112–2,004)
Old Babylonian (2,004–1,595 BCE)

By far most of the evidence I bring to bear concerns the period from 4,000 until 2,000 BCE, as it is both the key period of state formation and the focus of the bulk of the existing scholarship.

From time to time, I refer briefly to other early states, such as the Qin and Han dynasties of China, early Egypt, classical Greece, the Roman Republic and Empire, and even early Mayan civilization in the New World. The purpose of such excursions is to triangulate where the evidence from Mesopotamia is thin or disputed in order to make some educated guesses about patterns on the basis of comparisons. This is especially the case for the role of unfree labor in early states, the importance of disease in state collapse, the consequences of collapse, and, finally, the relationship between states and their "barbarians."

In explaining the surprises that awaited me and, I imagine, await my readers as well, I have relied on a large number of trusted "native trackers" in disciplinary territories with which I am not intimately familiar. The question is not whether I am poaching; I *mean* to poach! The question is rather whether I have poached from the most experienced, careful, well-traveled, and trusted native trackers. I will name some of my most important guides here because I do wish to implicate them in this enterprise insofar as their wisdom has helped me find my way. At the top of the list are archaeologists and specialists on the Mesopotamian alluvium who have been exceptionally generous with their time and critical advice: Jennifer Pournelle, Norman Yoffee, David Wengrow, and Seth Richardson. Others whose work has inspired me are, in no particular order: John McNeill, Edward Melillo, Melinda Zeder, Hans Nissen, Les Groube, Guillermo Algaze, Ann Porter, Susan Pollock, Dorian Q. Fuller, Andrea Seri, Tate Paulette, Robert Mc. Adams, Michael Dietler, Gordon Hillman, Karl Jacoby, Helen Leach, Peter Perdue, Christopher Beckwith, Cyprian Broodbank, Owen Lattimore, Thomas Barfield, Ian Hodder, Joe Manning, K. Sivaramakrishnan, Edward Friedman, Douglas Storm, James Prosek, Aniket Aga, Sarah Osterhoudt, Padraic Kenney, Gardiner Bovingdon, Timothy Pechora, Stuart Schwartz, Anna Tsing, David Graeber, Magnus Fiskesjo, Victor Lieberman, Wang Haicheng, Helen Siu, Bennet Bronson, Alex Lichtenstein, Cathy Shufro, Jeffrey Isaac, and Adam T. Smith. I am particularly grateful to Richard Manning, who, I found, anticipated a good part of my argument about cereal grains and states and whose intellectual large-spiritedness extended to allowing me to poach his title, *Against the Grain*, as the first element of my own.

Though not a little intimidated at the prospect, I tried out my arguments before audiences of archaeologists and specialists in ancient history. I want to thank them for their forbearance and helpful criticism. One of the first audiences on which I inflicted early revisions included many of my ex-colleagues at the University of Wisconsin, where I gave the Hilldale Lecture in 2013. I want also to thank Clifford Ando and his colleagues for inviting me to a conference on "Infrastructural and Despotic Power in Ancient States" at the University of Chicago in 2014, and David Wengrow and Sue Hamilton for the opportunity to give the V. Gordon Childe Lecture at the Institute of Archaeology, London, in 2016. Portions of my argument have been presented (and dissected!) at the University of Utah (the O. Meredith Wilson Lecture), the University of London's School of Oriental and African Studies (Centennial Lecture), Indiana University (Patten Lectures), the University of Connecticut, Northwestern University, the University of Frankfurt am Main, the Free University in Berlin, Columbia University's Legal Theory Workshop, and Aarhus University, which also afforded me the luxury of a paid leave during further researching and writing. I am especially grateful to my Danish colleagues Nils Bubandt, Mikael Gravers, Christian Lund, Niels Brimnes, Preben Kaarlsholm, and Bodil Frederickson for their intellectual generosity and for insights that contributed to my further education.

I don't believe anyone, anywhere ever had a more valuable and intellectually ferocious research assistant than I had in Annikki Herranan, now launched in her career as an anthropologist. Annikki laid out, week after week, an intellectual "tasting menu" of sumptuous proportions with an infallible guide to the juiciest morsels. Faizah Zakariah tracked down the permissions for the images found here, and Bill Nelson skillfully crafted the maps, charts, and "histograms" meant to help orient the reader. Finally, my Yale University Press editor, Jean Thomson Black, explains my loyalty, and that of many other authors, to the Press; she is the standard of quality, attention, and efficiency we all wish were not so rare. When it came to making sure that the final manuscript was as free of error, infelicities, and contradictions as it could possibly be, the "enforcer" was Dan Heaton. His insistence on perfection was made enjoyable by his high spirits and humor. Readers should know that everything was done to ensure that the remaining faults are irredeemably my own.

Yale Agrarian Studies Series
James C. Scott, Series Editor

The Agrarian Studies Series at Yale University Press seeks to publish outstanding and original interdisciplinary work on agriculture and rural society—for any period, in any location. Works of daring that question existing paradigms and fill abstract categories with the lived experience of rural people are especially encouraged.

—JAMES C. SCOTT, *Series Editor*

James C. Scott, *Seeing Like a State: How Certain Schemes to Improve the Human Condition Have Failed*

Steve Striffler, *Chicken: The Dangerous Transformation of America's Favorite Food*

Alissa Hamilton, *Squeezed: What You Don't Know About Orange Juice*

James C. Scott, *The Art of Not Being Governed: An Anarchist History of Upland Southeast Asia*

Sara M. Gregg, *Managing the Mountains: Land Use Planning, the New Deal, and the Creation of a Federal Landscape in Appalachia*

Michael R. Dove, *The Banana Tree at the Gate: A History of Marginal Peoples and Global Markets in Borneo*

Edwin C. Hagenstein, Sara M. Gregg, and Brian Donahue, eds., *American Georgics: Writings on Farming, Culture, and the Land*

Timothy Pachirat, *Every Twelve Seconds: Industrialized Slaughter and the Politics of Sight*

Andrew Sluyter, *Black Ranching Frontiers: African Cattle Herders of the Atlantic World, 1500–1900*

Brian Gareau, *From Precaution to Profit: Contemporary Challenges to Environmental Protection in the Montreal Protocol*

Kuntala Lahiri-Dutt and Gopa Samanta, *Dancing with the River: People and Life on the Chars of South Asia*

Alon Tal, *All the Trees of the Forest: Israel's Woodlands from the Bible to the Present*

Felix Wemheuer, *Famine Politics in Maoist China and the Soviet Union*

Jenny Leigh Smith, *Works in Progress: Plans and Realities on Soviet Farms, 1930–1963*

Graeme Auld, *Constructing Private Governance: The Rise and Evolution of Forest, Coffee, and Fisheries Certification*

Jess Gilbert, *Planning Democracy: Agrarian Intellectuals and the Intended New Deal*

Jessica Barnes and Michael R. Dove, eds., *Climate Cultures: Anthropological Perspectives on Climate Change*

Shafqat Hussain, *Remoteness and Modernity: Transformation and Continuity in Northern Pakistan*

Edward Dallam Melillo, *Strangers on Familiar Soil: Rediscovering the Chile-California Connection*

Devra I. Jarvis, Toby Hodgkin, Anthony H. D. Brown, John Tuxill, Isabel López Noriega, Melinda Smale, and Bhuwon Sthapit, *Crop Genetic Diversity in the Field and on the Farm: Principles and Applications in Research Practices*

Nancy J. Jacobs, *Birders of Africa: History of a Network*

Catherine A. Corson, *Corridors of Power: The Politics of U.S. Environmental Aid to Madagascar*

Kathryn M. de Luna, *Collecting Food, Cultivating People: Subsistence and Society in Central Africa Through the Seventeenth Century*

Carl Death, *The Green State in Africa*

James C. Scott, *Against the Grain: A Deep History of the Earliest States*

For a complete list of titles in the Yale Agrarian Studies Series, visit yalebooks.com/agrarian.

Against the Grain

Introduction: A Narrative in Tatters: What I Didn't Know

ow did *Homo sapiens sapiens* come, so very recently in its species history, to live in crowded, sedentary communities packed with domesticated livestock and a handful of cereal grains, governed by the ancestors of what we now call states? This novel ecological and social complex became the template for virtually all of our species' recorded history. Vastly amplified by population growth, water and draft power, sailing ships and long-distance trade, this template prevailed for more than six millennia until the use of fossil fuels. The account that follows is animated by a curiosity about the origin, structure, and consequences of this fundamentally agrarian, ecological complex.

The narrative of this process has typically been told as one of progress, of civilization and public order, and of increasing health and leisure. Given what we now know, much of this narrative is wrong or seriously misleading. The purpose of this book is to call that narrative into question on the basis of my reading of the advances in archaeological and historical research over the past two decades.

The founding of the earliest agrarian societies and states in Mesopotamia occurred in the latest five percent of our history as a species on the planet. And by that metric, the fossil fuel era, beginning at the end of the eighteenth century, represents merely the last quarter of a percent of our species history. For reasons that are alarmingly obvious, we are increasingly preoccupied by our footprint on the earth's environment in this last era. Just how massive that impact has become is captured in the lively debate swirling around the term "Anthropocene," coined to name a new geological epoch during which the activities of humans became decisive in affecting the world's ecosystems and atmosphere.[1]

While there is no doubt about the decisive contemporary impact of human activity on the ecosphere, the question of *when* it became decisive is in dispute. Some propose dating it from the first nuclear tests, which deposited a permanent and detectable layer of radioactivity worldwide. Others propose starting the Anthropocene clock with the Industrial Revolution and the massive use of fossil fuels. A case could also be made for starting the clock when industrial society acquired the tools—for example, dynamite, bulldozers, reinforced concrete (especially for dams)—to radically alter the landscape. Of these three candidates, the Industrial Revolution is a mere two centuries old and the other two are still virtually within living memory. Measured by the roughly 200,000-year span of our species, then, the Anthropocene began only a few minutes ago.

I propose an alternative point of departure that is far deeper historically. Accepting the premise of an Anthropocene as a qualitative and quantitative leap in our environmental impact, I suggest that we begin with the use of fire, the first great hominid tool for landscaping—or, rather, niche construction. Evidence for the use of fire is dated at least 400,000 years ago and perhaps much earlier still, long predating the appearance of Homo sapiens.[2] Permanent settlement, agriculture, and pastoralism, appearing about 12,000 years ago, mark a further leap in our transformation of the landscape. If our concern is with the historical footprint of hominids, one might well identify a "thin" Anthropocene long before the more explosive and recent "thick" Anthropocene; "thin" largely because there were so very few hominids to wield these landscaping tools. Our numbers circa 10,000 BCE were a puny two million to four million worldwide, far less than a thousandth of our population today. The other decisive premodern invention was institutional: the state. The first states in the Mesopotamian alluvium pop up no earlier than about 6,000 years ago, several millennia after the first evidence of agriculture and sedentism in the region. No institution has done more to mobilize the technologies of landscape modification in its interest than the state.

A sense, then, for how we came to be sedentary, cereal-growing, livestock-rearing subjects governed by the novel institution we now call the state requires an excursion into deep history. History at its best, in my view, is the most subversive

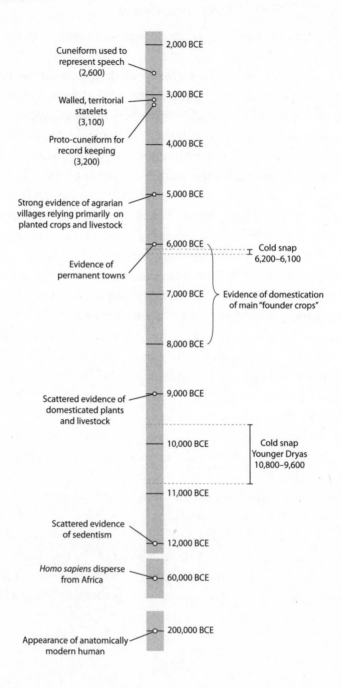

Cuneiform used to
represent speech
(2,600)

2,000 BCE

3,000 BCE

Walled, territorial
statelets
(3,100)

Proto-cuneiform for
record keeping
(3,200)

4,000 BCE

5,000 BCE

Strong evidence of agrarian
villages relying primarily on
planted crops and livestock

6,000 BCE

Cold snap
6,200–6,100

Evidence of
permanent towns

7,000 BCE

Evidence of domestication
of main "founder crops"

8,000 BCE

9,000 BCE

Scattered evidence of
domesticated plants
and livestock

10,000 BCE

Cold snap
Younger Dryas
10,800–9,600

11,000 BCE

Scattered evidence
of sedentism

12,000 BCE

Homo sapiens disperse
from Africa

60,000 BCE

200,000 BCE

Appearance of anatomically
modern human

discipline, inasmuch as it can tell us how things that we are likely to take for granted came to be. The allure of deep history is that by revealing the many contingencies that came together to shape, say, the Industrial Revolution, the Last Glacial Maximum, or the Qin Dynasty, it responds to the call by an earlier generation of French historians of the Annales School for a history of long-run processes (la longue durée) in place of a chronicle of public events. But the contemporary call for "deep history" goes the Annales School one better by calling for what often amounts to a species history. This is the zeitgeist in which I find myself, a zeitgeist surely illustrative of the maxim that "The Owl of Minerva flies only at dusk."[3]

PARADOXES OF STATE AND CIVILIZATION NARRATIVES

A foundational question underlying state formation is how we (*Homo sapiens sapiens*) came to live amid the unprecedented concentrations of domesticated plants, animals, and people that characterize states. From this wide-angle view, the state form is anything but natural or given. Homo sapiens appeared as a subspecies about 200,000 years ago and is found outside of Africa and the Levant no more than 60,000 years ago. The first evidence of cultivated plants and of sedentary communities appears roughly 12,000 years ago. Until then—that is to say for ninety-five percent of the human experience on earth—we lived in small, mobile, dispersed, relatively egalitarian, hunting-and-gathering bands. Still more remarkable,

---

Figure 1. (*opposite*) Timeline: From fire to cuneiform

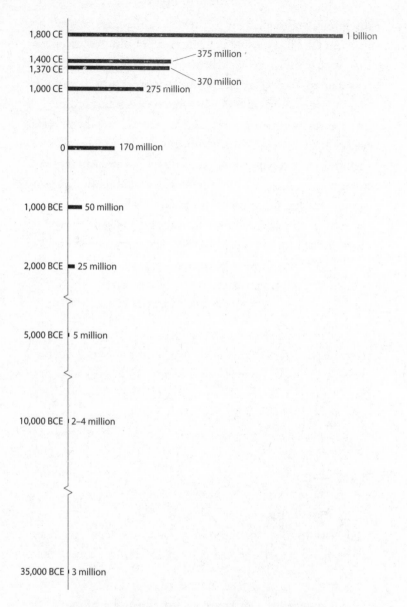

Figure 2. Estimated population in the ancient world

for those interested in the state form, is the fact that the very first small, stratified, tax-collecting, walled states pop up in the Tigris and Euphrates Valley only around 3,100 BCE, more than four millennia *after* the first crop domestications and sedentism. This massive lag is a problem for those theorists who would naturalize the state form and assume that once crops and sedentism, the technological and demographic requirements, respectively, for state formation were established, states/empires would immediately arise as the logical and most efficient units of political order.[4]

These raw facts trouble the version of human prehistory that most of us (I include myself here) have unreflectively inherited. Historical humankind has been mesmerized by the narrative of progress and civilization as codified by the first great agrarian kingdoms. As new and powerful societies, they were determined to distinguish themselves as sharply as possible from the populations from which they sprang and that still beckoned and threatened at their fringes. In its essentials, it was an "ascent of man" story. Agriculture, it held, replaced the savage, wild, primitive, lawless, and violent world of hunter-gatherers and nomads. Fixed-field crops, on the other hand, were the origin and guarantor of the settled life, of formal religion, of society, and of government by laws. Those who refused to take up agriculture did so out of ignorance or a refusal to adapt. In virtually all early agricultural settings the superiority of farming was underwritten by an elaborate mythology recounting how a powerful god or goddess entrusted the sacred grain to a chosen people.

Once the basic assumption of the superiority and attraction of fixed-field farming over all previous forms of subsistence is questioned, it becomes clear that this assumption itself rests on a deeper and more embedded assumption that is virtually never questioned. And that assumption is that sedentary life itself is superior to and more attractive than mobile forms of subsistence. The place of the domus and of fixed residence in the civilizational narrative is so deep as to be invisible; fish don't talk about water! It is simply assumed that weary Homo sapiens couldn't wait to finally settle down permanently, could not wait to end hundreds of millennia of mobility and seasonal movement. Yet there is massive evidence of determined resistance by mobile peoples everywhere to permanent settlement, even under relatively favorable circumstances. Pastoralists and hunting-and-gathering populations have fought against permanent settlement, associating it, often correctly, with disease and state control. Many Native American peoples were confined to reservations only on the heels of military defeat. Others seized historic opportunities presented by European contact to increase their mobility, the Sioux and Comanche becoming horseback hunters, traders, and raiders, and the Navajo becoming sheep-based pastoralists. Most peoples practicing mobile forms of subsistence — herding, foraging, hunting, marine collecting, and even shifting cultivation — while adapting to modern trade with alacrity, have bitterly fought permanent settlement. At the very least, we have no warrant at all for supposing that the sedentary "givens" of modern life can be read back into human history as a universal aspiration.[5]

The basic narrative of sedentism and agriculture has long survived the mythology that originally supplied its charter.

From Thomas Hobbes to John Locke to Giambattista Vico to Lewis Henry Morgan to Friedrich Engels to Herbert Spencer to Oswald Spengler to social Darwinist accounts of social evolution in general, the sequence of progress from hunting and gathering to nomadism to agriculture (and from band to village to town to city) was settled doctrine. Such views nearly mimicked Julius Caesar's evolutionary scheme from households to kindreds to tribes to peoples to the state (a people living under laws), wherein Rome was the apex, with the Celts and then the Germans ranged behind. Though they vary in details, such accounts record the march of civilization conveyed by most pedagogical routines and imprinted on the brains of schoolgirls and schoolboys throughout the world. The move from one mode of subsistence to the next is seen as sharp and definitive. No one, once shown the techniques of agriculture, would dream of remaining a nomad or forager. Each step is presumed to represent an epoch-making leap in mankind's well-being: more leisure, better nutrition, longer life expectancy, and, at long last, a settled life that promoted the household arts and the development of civilization. Dislodging this narrative from the world's imagination is well nigh impossible; the twelve-step recovery program required to accomplish that beggars the imagination. I nevertheless make a small start here.

It turns out that the greater part of what we might call the standard narrative has had to be abandoned once confronted with accumulating archaeological evidence. Contrary to earlier assumptions, hunters and gatherers—even today in the marginal refugia they inhabit—are nothing like the famished, one-day-away-from-starvation desperados of folklore. Hunters and gathers have, in fact, never looked so good—in terms of their diet, their health, and their leisure. Agriculturalists, on the contrary, have never looked so bad—in terms of *their* diet, *their* health, and *their* leisure.[6] The current fad of "Paleolithic" diets reflects the seepage of this archaeological knowledge into the popular culture. The shift from hunting and foraging to agriculture—a shift that was slow, halting, reversible, and sometimes incomplete—carried at least as many costs as benefits. Thus while the planting of crops has seemed, in the standard narrative, a crucial step toward a utopian present, it cannot have looked that way to those who first experienced it: a fact some scholars see reflected in the biblical story of Adam and Eve's expulsion from the Garden of Eden.

The wounds the standard narrative has suffered at the hands of recent research are, I believe, life threatening. For example, it has been assumed that fixed residence—sedentism— was a consequence of crop-field agriculture. Crops allowed populations to concentrate and settle, providing a necessary condition for state formation. Inconveniently for the narrative, sedentism is actually quite common in ecologically rich and varied, preagricultural settings—especially wetlands bordering the seasonal migration routes of fish, birds, and larger game. There, in ancient southern Mesopotamia (Greek for "between the rivers"), one encounters sedentary populations, even towns, of up to five thousand inhabitants with little or no agriculture. The opposite anomaly is also encountered: crop planting associated with mobility and dispersal except for a brief harvest period. This last paradox alerts us again to the fact that the implicit assumption of the standard narrative—namely that people couldn't wait to abandon mobility altogether and "settle down"—may also be mistaken.

Perhaps most troubling of all, the civilizational act at the center of the entire narrative: *domestication* turns out to be stubbornly elusive. Hominids have, after all, been shaping the plant world—largely with fire—since before Homo sapiens. What counts as the Rubicon of domestication? Is it tending wild plants, weeding them, moving them to a new spot, broadcasting a handful of seeds on rich silt, depositing a seed or two in a depression made with a dibble stick, or ploughing? There appears to be no "aha!" or "Edison light bulb" moment. There are, even today, large stands of wild wheat in Anatolia from which, as Jack Harlan famously showed, one could gather enough grain with a flint sickle in three weeks to feed a family for a year. Long before the deliberate planting of seeds in ploughed fields, foragers had developed all the harvest tools, winnowing baskets, grindstones, and mortars and pestles to process wild grains and pulses.[7] For the layman, dropping seeds in a prepared trench or hole seems decisive. Does discarding the stones of an edible fruit into a patch of waste vegetable compost near one's camp, knowing that many will sprout and thrive, count?

For archaeo-botanists, evidence of domesticated grains depended on finding grains with nonbrittle rachis (favored intentionally and unintentionally by early planters because the seedheads did not shatter but "waited for the harvester") and larger seeds. It now turns out that these morphological changes seem to have occurred well *after* grain crops had been cultivated. What had appeared previously to be unambiguous skeletal evidence of fully domesticated sheep and goats has also been called into question. The result of these ambiguities is twofold. First, it makes the identification of a single domestication event both arbitrary and pointless. Second, it reinforces the case for a very, very long period of what some have called "low-level food production" of plants not entirely wild and yet not fully domesticated either. The best analyses of plant domestication abolish the notion of a singular domestication event and instead argue, on the basis of strong genetic and archaeological evidence, for processes of cultivation lasting up to three millennia in many areas and leading to multiple, scattered domestications of most major crops (wheat, barley, rice, chick peas, lentils).[8]

While these archaeological findings leave the standard civilizational narrative in shreds, one can perhaps see this early period as part of a long process, still continuing, in which we humans have intervened to gain more control over the reproductive functions of the plants and animals that interest us. We selectively breed, protect, and exploit them. One might arguably extend this argument to the early agrarian states and their patriarchal control over the reproduction of women, captives, and slaves. Guillermo Algaze puts the matter even more boldly: "Early Near Eastern villages domesticated plants and animals. Uruk urban institutions, in turn, domesticated humans."[9]

PUTTING THE STATE IN ITS PLACE

Any inquiry into state formation like this one risks, by definition, giving the state a place of privilege greater than it might otherwise merit in a more balanced account of human affairs.

I wish to avoid this. The facts as I have come to understand them are that an evenhanded species history would give the state a far more modest role than it is normally accorded.

That states would have come to dominate the archaeological and historical record is no mystery. For us—that is to say Homo sapiens—accustomed to thinking in units of one or a few lifetimes, the permanence of the state and its administered space seems an inescapable constant of our condition. Aside from the utter hegemony of the state form today, a great deal of archaeology and history throughout the world is state-sponsored and often amounts to a narcissistic exercise in self-portraiture. Compounding this institutional bias is the archaeological tradition, until quite recently, of excavation and analysis of major historical ruins. Thus if you built, monumentally, in stone and left your debris conveniently in a single place, you were likely to be "discovered" and to dominate the pages of ancient history. If, on the other hand, you built with wood, bamboo, or reeds, you were much less likely to appear in the archaeological record. And if you were hunter-gatherers or nomads, however numerous, spreading your biodegradable trash thinly across the landscape, you were likely to vanish entirely from the archaeological record.

Once written documents—say, hieroglyphics or cuneiform—appear in the historical record, the bias becomes even more pronounced. These are invariably state-centric texts: taxes, work units, tribute lists, royal genealogies, founding myths, laws. There are no contending voices, and efforts to read such texts against the grain are both heroic and exceptionally difficult.[10] The larger the state archives left behind, generally speaking, the more pages devoted to that historical kingdom and its self-portrait.

And yet the very first states to appear in the alluvial and wind-blown silt in southern Mesopotamia, Egypt, and the Yellow River were minuscule affairs both demographically and geographically. They were a mere smudge on the map of the ancient world and not much more than a rounding error in a total global population estimated at roughly twenty-five million in the year 2,000 BCE. They were tiny nodes of power surrounded by a vast landscape inhabited by nonstate peoples—aka "barbarians." Sumer, Akkad, Egypt, Mycenae, Olmec/Maya, Harrapan, Qin China notwithstanding, most of the world's population continued to live outside the immediate grasp of states and their taxes for a very long time. When, precisely, the political landscape becomes definitively state-dominated is hard to say and fairly arbitrary. On a generous reading, until the past four hundred years, one-third of the globe was still occupied by hunter-gatherers, shifting cultivators, pastoralists, and independent horticulturalists, while states, being essentially agrarian, were confined largely to that small portion of the globe suitable for cultivation. Much of the world's population might never have met that hallmark of the state: a tax collector. Many, perhaps a majority, were able to move in and out of state space and to shift modes of subsistence; they had a sporting chance of evading the heavy hand of the state. If, then, we locate the era of definitive state hegemony as beginning about 1600 CE, the state can be said to dominate only the last two-tenths of one percent of our species' political life.

In focusing our attention on the exceptional places where

the earliest states appeared, we risk missing the key fact that in much of the world there was no state at all until quite recently. The classical states of Southeast Asia are roughly contemporaneous with Charlemagne's reign, more than six thousand years after the "invention" of farming. Those of the New World, with the exception of the Mayan Empire, are even more recent creations. They too were territorially quite small. Outside their reach were great congeries of "unadministered" peoples assembled in what historians might call tribes, chiefdoms, and bands. They inhabited zones of no sovereignty or vanishingly weak, nominal sovereignty.

The states in question were only rarely and then quite briefly the formidable Leviathans that a description of their most powerful reign tends to convey. In most cases, interregna, fragmentation, and "dark ages" were more common than consolidated, effective rule. Here again, we—and the historians as well—are likely to be mesmerized by the records of a dynasty's founding or its classical period, while periods of disintegration and disorder leave little or nothing in the way of records. Greece's four-century-long "Dark Age," when literacy was apparently lost, is nearly a blank page compared with the vast literature on the plays and philosophy of the Classical Age. This is entirely understandable if the purpose of a history is to examine the cultural achievements that we revere, but it overlooks the brittleness and fragility of state forms. In a good part of the world, the state, even when it was robust, was a seasonal institution. Until very recently, during the annual monsoon rains in Southeast Asia, the state's ability to project its power shrank back virtually to its palace walls. Despite the state's self-image and its centrality in most standard histories, it is important to recognize that for thousands of years after its first appearance, it was not a constant but a variable, and a very wobbly one at that in the life of much of humanity.

This is a nonstate history in yet another sense. It draws our attention to all those aspects of state making and state collapse that are either absent or leave only faint traces. Despite enormous progress in documenting climate change, demographic shifts, soil quality, and dietary habits, there are many aspects of the earliest states that one is unlikely to find chronicled in physical remains or in early texts because they are insidious, slow processes, perhaps symbolically threatening, and even unworthy of mention. For example, it appears that flight from the early state domains to the periphery was quite common, but, as it contradicts the narrative of the state as a civilizing benefactor of its subjects, it is relegated to obscure legal codes. I and others are virtually certain that disease was a major factor in the fragility of the early states. Its effects, however, are hard to document, since they were so sudden and so little understood, and because many epidemic diseases left no obvious bone signature. Similarly, the extent of slavery, bondage, and forced resettlement is hard to document as, in the absence of shackles, slave and free-subject remains are indistinguishable. All states were surrounded by nonstate peoples, but owing to their dispersal, we know precious little about their coming and going, their shifting relationship to states, and their political structures. When a city is burned to the ground, it is often hard to tell whether it was an accidental fire such as plagued all ancient cities built of combustible materials, a civil war or uprising, or a raid from outside.

To the degree that it is possible, I have tried to avert my gaze from the glare of state self-representation and have probed for historical forces systematically overlooked by dynastic and written histories and resistant to standard archaeological techniques.

THUMBNAIL ITINERARY

The theme of the first chapter turns on the domestication of fire, plants, and animals and the concentration of food and population such domestication makes possible. Before we could be made the object of state making, it was necessary that we gather—or *be* gathered—in substantial numbers with a reasonable expectation of not immediately starving. Each of these domestications rearranged the natural world in a way that vastly reduced the radius of a meal. Fire, which we owe to our older relative *Homo erectus*, has been our great trump card, allowing us to resculpt the landscape so as to encourage food-bearing plants—nut and fruit trees, berry bushes—and to create browse that would attract desirable prey. In cooking, fire rendered a host of previously indigestible plants both palatable and more nutritious. We owe our relatively large brain and relatively small gut (compared with other mammals, including primates), it is claimed, to the external predigestive help that cooking provides.

The domestication of grains—especially wheat and barley, in this case—and legumes furthers the process of concentration. Coevolving with humans, cultivars were selected especially for their large fruit (seeds), for their determinate ripening, and for their threshability (nonshattering quality).

They can be planted annually around the domus (the farmstead and its immediate surroundings) and provide a fairly reliable source of calories and protein—either as a reserve in a bad year or as a basic staple. Domesticated animals—especially sheep and goats, in this case—can be seen in the same light. They are our dedicated, four-footed (or, in the cases of chickens, ducks, and geese, two-footed) servant foragers. Thanks to their gut bacteria, they can digest plants that we cannot find and/or break down and can bring them back to us, as it were, in their "cooked" form as fat and protein, which we both crave and can digest. We selectively breed these domesticates for the qualities we desire: rapid reproduction, toleration of confinement, docility, meat, and milk and wool production.

The domestication of plants and animals was, as I have noted, not strictly necessary to sedentism, but it did create the conditions for an unprecedented level of concentration of food and population, especially in the most favorable agroecological settings: rich flood plain or loess soils and perennial water. This is why I choose to call such locations *late-Neolithic multispecies resettlement camps.* It turns out that while it provides ideal conditions for state making, the late-Neolithic multispecies resettlement camp involved a lot more drudgery than hunting and gathering and was not at all good for your health. Why anyone not impelled by hunger, danger, or coercion would willingly give up hunting and foraging or pastoralism for full-time agriculture is hard to fathom.

The term "domesticate" is normally understood as an active verb taking a direct object, as in "Homo sapiens domesticated rice . . . domesticated sheep," and so on. This overlooks the active agency of domesticates. It is not so clear, for example, to what degree we domesticated the dog or the dog domesticated us. And what about the "commensals"—sparrows, mice, weevils, ticks, bedbugs—that were not invited to the resettlement camp but gate-crashed anyway, as they found the company and the food congenial. And what about the "domesticators in chief," Homo sapiens? Were not they domesticated in turn, strapped to the round of ploughing, planting, weeding, reaping, threshing, grinding, all on behalf of their favorite grains and tending to the daily needs of their livestock? It is almost a metaphysical question who is the servant of whom—at least until it comes time to eat.

The meaning of domestication for plants, man, and beast is explored in Chapter 2. I argue, as have others, that domestication ought to be understood in an expansive way, as the ongoing effort of Homo sapiens to shape the entire environment to its liking. Given our frail knowledge about how the natural world works, one might say that the effort has been more abundant in *unintended* consequences than in intended effects. While the *thick* Anthropocene is judged by some to have begun with worldwide deposit of radioactivity following the dropping of the first atomic bomb, there is what I have termed a "thin" Anthropocene that dates from the use of fire by Homo erectus roughly half a million years ago and extends up through clearances for agriculture and grazing and the resulting deforestation, and siltation. The impact and tempo of this early Anthropocene grows as the world's population swells to roughly twenty-five million in 2,000 BCE. There is no particular reason to insist on the label "Anthropocene"—a term both in vogue and in much dispute as I write—but there are many reasons to insist on the global environmental impact of the domestication of fire, plants, and grazing animals.

"Domestication" changed the genetic makeup and morphology of both crops and animals around the domus. The assemblage of plants, animals, and humans in agricultural settlements created a new and largely artificial environment in which Darwinian selection pressure worked to promote new adaptations. The new crops became "basketcases," which could not survive without our constant attentions and protection. Much the same was true for domesticated sheep and goats, which became smaller, more placid, less aware of their surroundings and less sexually dimorphic. I ask in this context whether it is likely that a similar process affected us. How were we also domesticated by the domus, by our confinement, by crowding, by our different patterns of physical activity and social organization? Finally, by comparing the life world of agriculture—strapped as it is to the metronome of a major cereal grain—with the life world of the hunter-gatherer, I make the case that the life of farming is comparatively far narrower experientially and, in both a cultural and a ritual sense, more impoverished.

The burdens of life for nonelites in the earliest states, the subject of Chapter 3, were considerable. The first, as noted above, was drudgery. There is no doubt that, with the possible exception of flood recession (*décrue*) agriculture, farming was far more onerous than hunting and gathering. As Ester Boserup and others have observed, there is no reason why a forager in most environments would shift to agriculture unless forced to by population pressure or some form of coercion.

A second great and unanticipated burden of agriculture was the direct epidemiological effect of concentration—not just of people but of livestock, crops, and the large suite of parasites that followed them to the domus or developed there. Diseases with which we are now familiar—measles, mumps, diphtheria, and other community acquired infections—appeared for the first time in the early states. It seems almost certain that a great many of the earliest states collapsed as a result of epidemics analogous to the Antonine plague and the plague of Justinian in the first millennium CE or the Black Death of the fourteenth century in Europe. Then there was another plague: the state plague of taxes in the form of grain, labor, and conscription over and above onerous agricultural work. How, in such circumstances, did the early state manage to assemble, hold, and augment its subject population? Some have even argued that state formation was possible *only* in settings where the population was hemmed in by desert, mountains, or a hostile periphery.[11]

Chapter 4 is devoted to what might be called the grain hypothesis. It is surely striking that virtually all classical states were based on grain, including millets. History records no cassava states, no sago, yam, taro, plantain, breadfruit, or sweet potato states. ("Banana republics" don't qualify!) My guess is that only grains are best suited to concentrated production, tax assessment, appropriation, cadastral surveys, storage, and rationing. On suitable soil wheat provides the agro-ecology for dense concentrations of human subjects.

In contrast the tuber cassava (aka manioc, yucca) grows below ground, requires little care, is easy to conceal, ripens in a year, and, most important, can safely be left in the ground and remain edible for two more years. If the state wants your cassava, it will have to come and dig up the tubers one by one, and then it has a cartload of little value and great weight if transported. If we were evaluating crops from the perspective of the premodern "tax man," the major grains (above all, irrigated rice) would be among the most preferred, and roots and tubers among the least preferred.

It follows, I think, that state formation becomes possible only when there are few alternatives to a diet dominated by domesticated grains. So long as subsistence is spread across several food webs, as it is for hunter-gatherers, swidden cultivators, marine foragers, and so on, a state is unlikely to arise, inasmuch as there is no readily assessable and accessible staple to serve as a basis for appropriation. One might imagine that ancient domesticated legumes, say—peas, soybeans, peanuts, or lentils, all of which are nutritious and can be dried for storage—might serve as a tax crop. The obstacle in this case is that most legumes are indeterminate crops that can be picked as long as they grow; they do not have a determinate harvest, something the tax man requires.

Some agro-ecological settings may be considered "pre-adapted" for concentrating grain fields and population, owing to rich silt and plentiful water, and these areas are in turn possible locations for state making. Such settings are perhaps necessary for early state making, but not sufficient. One could say that the state has an elective affinity for such locations. Contrary to some earlier assumptions, the state did not invent irrigation as a way of concentrating population, let alone crop domestication; both were the achievements of prestate peoples. What the state has often done, once established, however, is to maintain, amplify, and expand the agro-ecological setting that is the basis of its power by what we might call state landscaping. This has included repairing silted channels, digging new feeder canals, settling war captives on arable land, penalizing subjects who are not cultivating, clearing new fields, forbidding nontaxable subsistence activities such as swiddening and foraging, and trying to prevent the flight of its subjects.

There is, I believe, something of an agro-economic module that characterizes most of the early states. Whether the grain in question is wheat, barley, rice, or maize—the four crops that account, even today, for more than half of the world's caloric consumption—the patterns display a family resemblance. The early state strives to create a legible, measured, and fairly uniform landscape of taxable grain crops and to hold on this land a large population available for corvée labor, conscription, and, of course, grain production. For dozens of reasons, ecological, epidemiological, and political, the state often fails to achieve this aim, but this is, as it were, the steady glint in its eye.

An alert reader might at this point ask, what is a state anyway? I think of the polities of early Mesopotamia as gradually becoming states. That is, "stateness," in my view, is an institutional continuum, less an either/or proposition than a judgment of more or less. A polity with a king, specialized administrative staff, social hierarchy, a monumental center, city walls, and tax collection and distribution is certainly a "state" in the strong sense of the term. Such states come into existence in the last centuries of the fourth millennium BCE and seem to be well attested at the latest by the strong Ur III territorial polity in southern Mesopotamia around 2,100. Before that there were polities with substantial populations, commerce, artisans, and, it seems, town assemblies, but one could argue about the degree to which these characteristics would satisfy a strong definition of stateness.

As may already be obvious, the southern Mesopotamian alluvium is at the center of my geographical attention for the simple reason that it was here that the first small states arose. "Pristine" is the adjective normally used to describe them. While fixed settlements and domesticated grains can be found earlier elsewhere (for example, in Jericho, the Levant, and the "hilly flanks" east of the alluvium), they did not give rise to states. Mesopotamian state forms, in turn, influenced subsequent state-making practices in Egypt, in northern Mesopotamia, and even in the Indus Valley. For this reason, and aided by surviving clay cuneiform tablets and the prodigious scholarship on the area, I concentrate on Mesopotamian states. When parallels or contrasts are striking and apposite, I refer occasionally to early state making in north China, Crete, Greece, Rome, and Maya.

One might be tempted to say that states arise, when they do, in ecologically rich areas. This would be a misunderstanding. What is required is wealth in the form of an appropriable, measurable, dominant grain crop and a population growing it that can be easily administered and mobilized. Areas of great but diverse abundance such as wetlands, which offer dozens of subsistence options to a mobile population, because of their very illegibility and fugitive diversity, are not zones of successful state making. The logic of assessable and accessible crops and people applies as well to smaller-scale efforts at

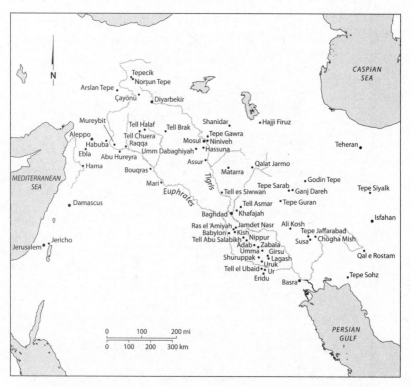

Figure 3. Mesopotamia: Tigris-Euphrates region

control and legibility one finds in the Spanish *redduciones* in the New World, many missionary settlements, and that paragon of legibility, the monocrop plantation with the workforce in the barracks.

The larger question, the one I address in Chapter 5, is important because it bears on the role of coercion in establishing and maintaining the ancient state. Though it is a subject of heated debate, the question goes directly to the heart of the traditional narrative of civilizational progress. If the forma-

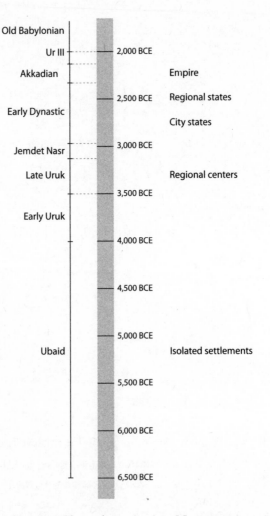

Figure 4. Chronology: Ancient Mesopotamia

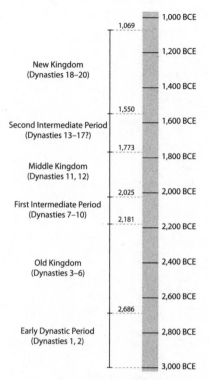

New Kingdom
(Dynasties 18–20)

1,069

1,000 BCE
1,200 BCE
1,400 BCE

1,550

Second Intermediate Period
(Dynasties 13–17?)

1,600 BCE

1,773

Middle Kingdom
(Dynasties 11, 12)

1,800 BCE

2,025

First Intermediate Period
(Dynasties 7–10)

2,000 BCE

2,181

2,200 BCE

Old Kingdom
(Dynasties 3–6)

2,400 BCE
2,600 BCE

2,686

Early Dynastic Period
(Dynasties 1, 2)

2,800 BCE
3,000 BCE

Figure 5. Chronology:
Ancient Nile River Egypt

tion of the earliest states were shown to be largely a coercive enterprise, the vision of the state, one dear to the heart of such social-contract theorists as Hobbes and Locke, as a magnet of civil peace, social order, and freedom from fear, drawing people in by its charisma, would have to be reexamined.

The early state, in fact, as we shall see, often failed to hold its population; it was exceptionally fragile epidemiologically, ecologically, and politically and prone to collapse or fragmen-

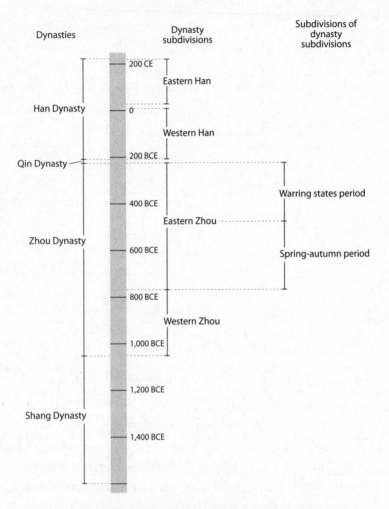

Figure 6. Chronology: Ancient Yellow River China

tation. If, however, the state often broke up, it was not for lack of exercising whatever coercive powers it could muster. Evidence for the extensive use of unfree labor—war captives, indentured servitude, temple slavery, slave markets, forced resettlement in labor colonies, convict labor, and communal slavery (for example, Sparta's helots)—is overwhelming. Unfree labor was particularly important in building city walls and roads, digging canals, mining, quarrying, logging, monumental construction, wool textile weaving, and of course agricultural labor. The attention to "husbanding" the subject population, including women, as a form of wealth, like livestock, in which fertility and high rates of reproduction were encouraged, is apparent. The ancient world clearly shared Aristotle's judgment that the slave was, like a plough animal, a "tool for work." Even before one encounters terms for slaves in the early written records, the archaeological record speaks volumes with its bas relief depictions of ragged captive slaves being led back from the field of victory and, in Mesopotamia, thousands of identical, small, beveled bowls used, in all likelihood, for barley or beer rations for gang labor.

Formal slavery in the ancient world reaches its apotheosis in classical Greece and early imperial Rome, which were slave states in the full sense one applies to the antebellum South in the United States. Chattel slavery on this order, though not absent in Mesopotamia and early Egypt, was less dominant than other forms of unfree labor, such as the thousands of women in large workshops in Ur making textiles for export. That a good share of the population in Greece and Roman Italy was being held against its will is testified to by slave rebellions in Roman Italy and Sicily, by the wartime offers of freedom—by Sparta to Athenian slaves and by the Athenians to Sparta's helots—and by the frequent references to fleeing and absconding populations in Mesopotamia. One is reminded in this context of Owen Lattimore's admonition that the great walls of China were built as much to keep Chinese taxpayers *in* as to keep the barbarians *out*. Variable as it is over time and hard as it is to quantify, bondage appears to have been a condition of the ancient state's survival. Early states surely did not invent the institution of slavery, but they did codify and organize it as a state project.

The earliest states were historically novel institutions; there were no manuals of statecraft, no Machiavelli rulers could consult, so it is not surprising that they were often short-lived. China's Qin Dynasty, famous for its many innovations of strong governance, lasted a mere fifteen years. The agro-ecology favorable to state making is relatively stationary, while the states that occasionally appear in these locations blink on and off like erratic traffic lights. The reasons for this fragility and how we might understand its larger meaning provide the theme of Chapter 6.

Much archaeological ink has been spilled trying to explain, for example, the Mayan "collapse," the Egyptian "First Intermediate Period," and Greece's "Dark Age." Frequently the evidence we have provides no dispositive clue. The causes are typically multiple, and it is arbitrary to single out one as decisive. As with a patient suffering many underlying illnesses, it is difficult to specify the cause of death. And when, say, a drought leads to hunger and then to resistance and flight of which, in turn, a neighboring kingdom takes advantage by invading, sacking the kingdom, and carrying off its population, which of these causes ought we to prefer? The sparse written record rarely helps. When a kingdom is destroyed by invasion, raids, civil war, or rebellion, the deposed scribes rarely remain at their posts long enough to record the debacle. Occasionally there is evidence that a palace complex has been burned—but by whom and for what reason is rarely clear.

Here, I emphasize particularly those causes of fragility that are intrinsic to the agro-ecology of the earliest states. Extrinsic causes—say, drought or climate change (which is clearly implicated in several regionwide simultaneous "collapses")— may in fact be more important overall in state collapse, but intrinsic causes tell us more about the self-limiting aspects of early states. To this end, I speculate on three fault lines that are by-products of state formation itself. The first are the disease effects of the unprecedented concentrations of crops, people, and livestock together with their attendant parasites and pathogens. I imagine, as do others, that epidemics of one kind or another, including crop diseases, were responsible for quite a few sudden collapses. Evidence, however, is difficult to come by. More insidious are two ecological effects of urbanism and intensive irrigated agriculture. The former resulted in steady deforestation of the upstream watershed of riverine states and subsequent siltation and floods. The latter resulted in well-documented salinization of the soil, lower yields, and eventual abandonment of arable land.

I want, finally, to question, as others have, the use of the term "collapse" to describe many of these events.[12] In unreflective use, "collapse" denotes the civilizational tragedy of a great early kingdom being brought low, along with its cultural achievements. We should pause before adopting this usage. Many kingdoms were, in fact, confederations of smaller settlements, and "collapse" might mean no more than that they have, once again, fragmented into their constituent parts, perhaps to reassemble later. In the case of reduced rainfall and crop yields, "collapse" might mean a fairly routine dispersal to deal with periodic climate variation. Even in the case of, say, flight or rebellion against taxes, corvée labor, or conscription, might we not celebrate—or at least not deplore—the destruction of an oppressive social order? Finally, in case it is the so-called barbarians who are at the gate, we should not forget that they often adopt the culture and language of the rulers whom they depose. Civilizations should never be confused with the states that they typically outlast, nor should we unreflectively prefer larger units of political order to smaller units.

And what about these barbarians who, in the epoch of the early states, are massively more numerous than state subjects and, though dispersed, occupy most of the earth's habitable surface? The term "barbarian," we know, was originally applied by the Greeks to all non-Greek speakers—captured slaves as well as quite "civilized" neighbors such as the Egyptians, the Persians, and the Phoenicians. "Ba-ba" was meant to be a parody of the sound of non-Greek speech. In one form or another the term was reinvented by all early states to distinguish themselves from those outside the state. It is fitting, therefore, that my seventh and last chapter is devoted to the "barbarians" who were simply the vast population not subject to state control. I will continue to use the term "barbarian"—with tongue planted firmly in cheek—in part because I want to argue that the era of the earliest and fragile states was a time when it was good to be a barbarian. The length of this period varied from place to place depending on state strength and military technology; while it lasted it might be called the golden age of barbarians. The barbarian zone, as it were, is essentially the mirror image of the agro-ecology of the state. It is a zone of hunting, slash-and-burn cultivation, shellfish collection, foraging, pastoralism, roots and tubers, and few if any standing grain crops. It is a zone of physical mobility, mixed and shifting subsistence strategies: in a word, "illegible" production. If the barbarian realm is one of diversity and complexity, the state realm is, agro-economically speaking, one of relative simplicity. Barbarians are not essentially a cultural category; they are a political category to designate populations not (yet?) administered by the state. The line on the frontier where the barbarians begin is that line where taxes and grain end. The Chinese used the terms "raw" and "cooked" to distinguish between barbarians. Among groups with the same language, culture, and kinship systems, the "cooked" or more "evolved" segment comprised those whose households had been registered and who were, however nominally, ruled by Chinese magistrates. They were said to "have entered the map."

As sedentary communities, the earliest states were vulnerable to more mobile nonstate peoples. If one thinks of hunters and foragers as specialists at locating and exploiting food sources, the static aggregations of people, grain, livestock, textiles, and metal goods of sedentary communities represented relatively easy pickings. Why should one go to the trouble of growing a crop when, like the state (!), one can simply confiscate it from the granary. As the Berber saying so eloquently attests, "Raiding is our agriculture." The growth of sedentary agricultural settlements that were everywhere the foundation of early states can be seen as a new and very lucrative foraging site for nonstate peoples—one-stop shopping, as it were. As Native Americans realized, the tame European cow was easier to "hunt" than the white-tailed deer. The consequences for the early state were considerable. Either it invested heavily in defenses against raiding and/or it paid tribute—protection money—to potential raiders in return for not plundering. In either case the fiscal burden on the early state, and hence its fragility, increased appreciably.

While raiding's spectacular quality tends to dominate accounts of the early state's relationship with barbarians, it was surely far less important than trade. The early states, located for the most part in rich, alluvial bottomlands, were natural trading partners with nearby barbarians. Ranging widely in a far more diverse environment, only the barbarians could supply the necessities without which the early state could not long survive: metal ores, timber, hides, obsidian, honey, medicinals, and aromatics. The lowland kingdom was more valuable as a trade depot, in the long run, than as a site of plunder. It represented a large, new, and lucrative market for products from the hinterland that could be traded for lowland products such as grain, textiles, dates, and dried fish. Once the development of coastal shipping allowed for more long-distance trade, the volume of this trade exploded. To imagine the effect one need only think of the impact the market for beaver pelts in Europe had on Native American hunting. Both foraging and hunting became, with the expansion of trade, more a trading and entrepreneurial venture than a pure subsistence activity.

The result of this symbiosis was a cultural hybridity far greater than the typical "civilized-barbarian" dichotomy would allow. A convincing case has been made that the early state or empire was usually shadowed by a "barbarian twin," which rose with it and shared its fate when it fell.[13] The Celtic trading *oppida* at the fringe of the Roman Empire provide an example of this dependency.

Thus the long era of relatively weak agrarian states and numerous, mounted, nonstate peoples was something of a golden age of barbarians; they enjoyed a profitable trade with the early states, augmented with tribute and raiding when necessary; they avoided the inconveniences of taxes and agricultural labor; they enjoyed a more nutritious and varied diet and greater physical mobility.

Two aspects of this trade, however, were both melancholy and fateful. Perhaps the main commodity traded to the early states was the slave—typically from among the barbarians. The ancient states replenished their population by wars of capture and by buying slaves on a large scale from barbarians who specialized in the trade. In addition, it was a rare early state that did not engage barbarian mercenaries for its defense. Selling both their fellow barbarians and their martial service to the early states, the barbarians contributed mightily to the decline of their brief golden age.

CHAPTER ONE

The Domestication of Fire, Plants, Animals, and . . . Us

FIRE

WHAT fire meant for hominids and ultimately for the rest of the natural world is presaged vividly by a cave excavation in South Africa.[1] At the deepest and therefore oldest strata, there are no carbon deposits and hence no fire. Here one finds full skeletal remains of large cats and fragmentary bone shards—bearing tooth marks—of many fauna, among which is Homo erectus. At a higher, later stratum, one finds carbon deposits signifying fire. Here, there are full skeletal remains of Homo erectus and fragmentary bone shards of various mammals, reptiles, and birds, among which are a few gnawed bones of large cats. The change in cave "ownership" and the reversal in who was apparently eating whom testify eloquently to the power of fire for the species that first learned to use it. At the very least, fire provided warmth, light, and relative safety from nocturnal predators as well as a precursor to the domus or hearth.

The case for the use of fire being *the* decisive transformation in the fortunes of hominids is convincing. It has been mankind's oldest and greatest tool for reshaping the natural world. "Tool," however, is not quite the right word; unlike an inanimate knife, fire has a life of its own. It is, at best, a "semidomesticate," appearing unbidden and, if not guarded carefully, escaping its shackles to become dangerously feral.

Hominids' use of fire is historically deep and pervasive. Evidence for human fires is at least 400,000 years old, long before our species appeared on the scene. Thanks to hominids, much of the world's flora and fauna consist of fire-adapted species (pyrophytes) that have been encouraged by burning. The effects of anthropogenic fire are so massive that they might be judged, in an evenhanded account of the human impact on the natural world, to overwhelm crop and livestock domestications. Why human fire as landscape architect doesn't register as it ought to in our historical accounts is perhaps that its effects were spread over hundreds of millennia and were accomplished by "precivilized" peoples also known as "savages." In our age of dynamite and bulldozers, it was a very slow-motion sort of environmental landscaping. But its aggregate effects were momentous.

Our ancestors could not have failed to notice how natural wildfires transformed the landscape: how they cleared old vegetation and encouraged a host of quick-colonizing grasses and shrubs, many bearing desired seeds, berries, fruits, and nuts. They could also not have failed to notice that a fire drove fleeing game from its path, exposed hidden burrows and nests of small game, and, most important, later stimulated the browse and mushrooms that attracted grazing prey. Native North Americans deployed fire to sculpt landscapes favored by elk, deer, beaver, hare, porcupine, ruffed grouse, turkey, and quail, all of which they hunted. The game they subsequently bagged represented a kind of *harvesting* of prey animals they had deliberately assembled by carefully creating a habitat they would find enticing.[2] Quite apart from being the designers of hunting grounds—veritable game parks—early humans used fire to hunt large game. The evidence suggests that long before the bow and arrow appeared, roughly twenty thousand years ago, hominids were using fire to drive herd animals off precipices and to drive elephants into bogs where, immobilized, they could more easily be killed.

Fire was the key to humankind's growing sway over the natural world—a species monopoly and trump card, worldwide. The Amazonian rain forest bears indelible traces of the use of fire to clear land and open the canopy; Australia's eucalyptus landscape is, to a considerable degree, the effect of human fire. The volume of such landscaping in North America was such that when it stopped abruptly, due to the devastating epidemics that came with the European, the newly unchecked growth of forest cover created the illusion among white settlers that North America was a virtually untouched, primeval forest. According to some climatologists, the cold spell known as the Little Ice Age, from roughly 1500 to 1850, may well have been due to the reduction of CO_2—a greenhouse gas—brought about by the die-off of North America's indigenous fire farmers.[3]

From our perspective, what this slow-motion landscape engineering accomplishes over time is to concentrate more subsistence resources in a smaller and smaller area. It rearranges, by a fire-assisted form of applied horticulture, desirable flora and fauna in a tighter ring around the camp(s) and makes hunting and forging easier. The radius of a meal, one might say, is reduced. Subsistence resources are closer at hand, more abundant, and more predictable. Wherever humans and fire were at work sculpting the landscape for hunting-and-gathering convenience, few nutrient-poor "climax" forests were allowed to develop. We are nowhere near the oxen, the plough, and the tame livestock of the domus, but we are nevertheless looking at a systematic intensification of landscape and resource management of massive proportions that predates by hundreds of millennia the actual cultivation of fully domesticated crops and pastoralism. Unlike optimal foraging theory that takes the disposition of the natural world as given and asks how a rational actor would distribute his or her efforts in procuring food, what we have here is a deliberate disturbance ecology in which hominids create, over time, a mosaic of biodiversity and a distribution of desirable resources more to their liking. Evolutionary biologists term such activity, combining location, repositioning of resources, and physical safety, niche construction: think "beaver." Seeing the concentration of resources in this light places the milestones of the classical civilizational narrative—the domestication of plants and animals—in a new light as elements among many in a longue-durée continuum of ever-more-elaborate niche construction.[4]

Fire powerfully concentrates people in yet another way: cooking. It is virtually impossible to exaggerate the importance of cooking in human evolution. The application of fire to raw food externalizes the digestive process; it gelatinizes starch and denatures protein. The chemical disassembly of raw food, which in a chimpanzee requires a gut roughly three times the size of ours, allows Homo sapiens to eat far less food and expend far fewer calories extracting nutrition from it. The effects are enormous. It allowed early man to gather and eat a far wider range of foods than before: plants with thorns, thick skins, and bark could be opened, peeled, and detoxified by cooking; hard seeds and fibrous foods that would not have repaid the caloric costs of digesting them became palatable; the flesh and guts of small birds and rodents could be sterilized. Even before the advent of cooking, Homo sapiens was a broad-spectrum omnivore, pounding, grinding, mashing, fermenting, and pickling raw meat and plants, but with fire, the range of foods she could digest expanded exponentially. As testimony to that range, an archaeological site in the Rift Valley dated twenty-three thousand years ago gives evidence of a diet spanning four food webs (water, woodland, grassland, and arid) encompassing at least 20 large and small animals, 16 families of birds, and 140 kinds of fruit, nuts, seeds, and pulses, not to mention plants for medicinal and craft purposes—baskets, weaving, traps, weirs.[5]

Fire for cooking was at least as important as fire as landscape architect for the concentration of population. The latter placed more desirable foods within easier reach, while the former rendered a whole range of hitherto indigestible foods now both nutritious and palatable. The radius of a meal was much further reduced. Not only that, but softer cooked foods as a form of external premastication allowed easier weaning and the feeding of the elderly and toothless.

Armed with fire to sculpt the environment and able to eat so much more of it, early man could both stay closer to the hearth and, at the same time, establish new hearths in previously forbidding environments. Neanderthal colonization of northern Europe is a case in point; it would have been inconceivable without fire for warmth, hunting, and cooking.

The genetic and physiological effects of at least half a million years of cooking have been enormous. Compared with our primate cousins, we have a gut less than half the size and far smaller teeth, and we spend far fewer calories chewing and digesting. The gains in nutritional efficiency, Richard Wrangham claims, largely account for the fact that our brains are three times the size one would expect, judging by other mammals.[6] In the archaeological record the surge in brain size coincides with hearths and the remains of meals. Morphological changes of this magnitude have been known to occur in other animals in as little as twenty thousand years following a dramatic shift in diet and ecological niche.

Fire largely accounts for our reproductive success as the world's most successful "invasive."[7] Much like certain trees, plants, and fungi, we are a fire-adapted species: pyrophytes. We have adapted our habits, diet, and body to the characteristics of fire, and having done so, we are chained, as it were, to its care and feeding. If the litmus test of domestication for a plant or animal is that it cannot propagate itself without our assistance, then, by the same token, we have adapted so massively to fire that our species would have no future without it. Even overlooking entirely the fire-dependent crafts that developed later—potter, blacksmith, baker, brick maker, glassmaker, metalworker, gold- and silversmith, brewer, charcoal maker, food smoker, plaster maker—it is no exaggeration to say that we are utterly dependent on fire. It has in a real sense domesticated us. One small but telling piece of evidence is that raw-foodists who insist on cooking nothing invariably lose weight.[8]

CONCENTRATION AND SEDENTISM:
A WETLANDS THESIS

What might have been an earlier trend toward population growth and settlement in the Fertile Crescent owing to warmer and wetter conditions ended abruptly around 10,800 BCE. A millennium-long cold snap that followed is believed by some to have been caused by a massive pulse of glacial melt from North America (Lake Agassiz) suddenly draining eastward into the Atlantic through what we now call the Saint Lawrence River.[9] Population receded, the remainder shrank back from marginal highlands to refugia where the climate, and therefore the flora and fauna, were more favorable. Then, around 9,600 BCE, the cold snap broke and it became warmer and wetter again—and fast. The average temperature may have increased as much as seven degrees Celsius within a single decade. The trees, mammals, and birds burst out of the refugia to colonize a suddenly more hospitable landscape—and with them, of course, their companion species, Homo sapiens.

At about the same time, archaeologists find scattered evidence of yearlong occupation of many sites—the Natufian Period in the southern Levant and the "prepottery" stage in Neolithic villages in Syria, central Turkey, and western Iran. They generally occur in water-rich areas and subsist largely by hunting and foraging, though there is evidence—disputed—of cereal horticulture and livestock rearing. Not in dispute, however, is that between 8,000 and 6,000 BCE, all the so-called "founder crops"—the cereals and legumes: lentils, peas, chickpeas, bitter vetch, and flax (for cloth)—are being planted, though generally on a modest scale. Over the same two-millennia span—the timing vis-à-vis cereals is not clear—domesticated goats, sheep, pigs, and cattle make their appearance. With this suite of domesticates the full "Neolithic package," seen as the decisive agricultural revolution that marks the beginning of civilization, including the first small urban agglomerations, is in place.

Permanent proto-urban settlements emerge in the wetlands of the southern alluvium near the Persian Gulf around 6,500 BCE. The southern alluvium is not the earliest site of year-round settlements; nor is it the site where the first evidence of domesticated cereals appears. In these respects, it is a latecomer. I concentrate in this book on these later sites for two important reasons. First, these urban agglomerations at the mouth of the Euphrates—for example, Eridu, Ur, Umma, and Uruk—go on to become, much later, the very first "statelets" in the world. Second, while other ancient societies such as Egypt, the Levant, the Indus Valley, the Yellow River Valley, and Maya in the New World have their own variants of the Neolithic revolution, southern Mesopotamia not only was the site of the first state system, but it also directly influenced later state making elsewhere in the Middle East as well as in Egypt and India.

Even on the basis of this rough-and-ready chronology—much of it still in some dispute—one can see how much of it is stubbornly at odds with what I have called the standard civilizational narrative. That narrative pivoted on the domestication of grain as the basic precondition of permanent

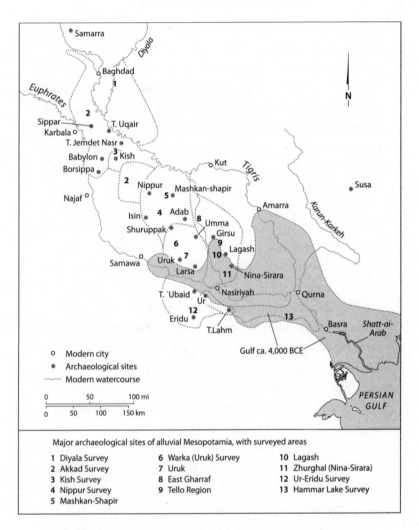

Figure 7. Mesopotamian alluvium: Archaeological sites

sedentary life, and thus of towns, cities, and civilization. The presumption, still commonly held, was that hunting and foraging required such mobility and dispersal that sedentism was out of the question. Yet sedentism long predates the domestication of grains and livestock and often persists in settings where there is little or no cereal cultivation. What is also absolutely clear is that domesticated grains and livestock are known long before anything like an agrarian state appears—far longer than previously imagined. On the basis of the latest evidence, the gap between these two key domestications and the first agrarian economies based on them is now reckoned to stretch for 4,000 years.[10] Clearly our ancestors did not rush headlong into the Neolithic revolution or into the arms of the earliest states.

Those who crafted the older narrative were radically mistaken in yet another respect. Taking as their point of departure the exceptionally arid conditions that have prevailed in the Tigris-Euphrates Valley in recent history, they reasonably enough projected this aridity back to the dawn of agriculture. Confined in limited oases and river valleys, a growing population was assumed to have been obliged to intensify its subsistence practices in order to extract more from limited arable land. The only viable intensification strategy was irrigation, for which there was archaeological evidence. Irrigation alone could guarantee the abundant harvests where rainfall was so woefully inadequate. In turn, such a huge project of landscape modification required the mobilization of labor to dig and maintain the canals, which implied the existence of a public authority capable of assembling and disciplining that labor force. Irrigation works made for a dense agro-pastoral economy that, they assumed, fostered state formation as a condition of its existence.

WETLANDS AND SEDENTISM

The prevailing view that "making the desert bloom" by irrigated agriculture was the foundation of the first substantial sedentary communities, however, turns out to be mistaken in nearly every particular. As we shall see, the earliest large fixed settlements sprang up in wetlands, not arid settings; they relied overwhelmingly on wetland resources, not grain, for their subsistence; and they had no need of irrigation in the generally understood sense of the term. Insofar as any human landscaping was necessary in this setting, it was far more likely to be drainage than irrigation. The classical view that ancient Sumer was a miracle of irrigation organized by the state in an arid landscape turns out to be totally wrong. We owe the most comprehensive and documented revisionist case along these lines to Jennifer Pournelle's pathbreaking study of the southern Mesopotamian alluvium during the seventh and sixth millennia BCE.[11]

Southern Mesopotamia at that time was not at all arid, but rather more like a foragers' wetland paradise. Owing to the substantial rise in sea levels and the flatness of the Tigris-Euphrates delta, there was a massive marine "transgression" into areas that are now arid. Pournelle reconstructs this vast deltaic wetland zone on the basis of remote sensing, earlier aerial surveys, hydrological history, readings of ancient sediments and water courses, climate history, and archaeological remains. The mistake of most (not all) earlier observers had

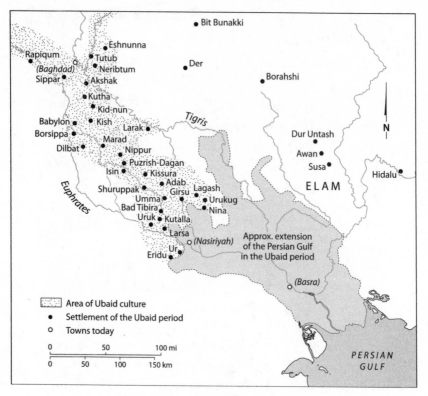

Figure 8. Mesopotamian alluvium: Persian Gulf extension, circa 6,500 BCE. Courtesy of Jennifer Pournelle

not only been to project the general aridity of the region back ten thousand years but also to ignore the fact that the alluvium was then—before the annual depositions of sediment—more than ten meters below its current level. The waters of the Persian Gulf, under those earlier conditions, lapped at the door of ancient Ur, now quite far inland, and tidal salt water extended northward as far as Nasiriya and Amara.

A brief description of how substantial populations depending largely on wild, free-living plant and marine resources could arise without benefit of irrigation of substantial cereal crops will illuminate two issues of analytical concern. First, it demonstrates the stability and richness of a subsistence based on several diverse food webs. Much of the diet during the Ubaid Period (6,500–3,800 BCE, named for a widespread pottery style) came from fish, birds, and turtles that teemed in the wetlands. Second, it will later serve to show how the very breadth of a subsistence web—hunting, fishing, foraging, and gathering in a variety of ecological settings—poses insurmountable obstacles to the imposition of a single political authority.

Rather than an arid zone between two rivers, as it largely is today, the southern alluvium was an intricate deltaic wetland crisscrossed by hundreds of distributaries, now merging, now diverging, with each season of flooding. The alluvium operated as a great sponge, absorbing the annual high water flow, raising the water table, then releasing it slowly in the dry months beginning in May. The flood plain of the lower Euphrates is extremely flat: the gradient varies from twenty to thirty centimeters per kilometer in the north to a mere two to three centimeters per kilometer in the south, making the river's historical course highly erratic.[12] At the height of the annual flooding the water courses regularly overtopped their natural ridges or levees, created by the annual deposition of their coarser sediments, and spilled down the backslope, flooding the adjacent lowlands and depressions. As the beds of many watercourses were above the surrounding land, a simple breach in the levee at high water would accomplish the same purpose—we might call this last technique "assisted natural irrigation."[13] Seed grains could be broadcast on the naturally prepared field. The nutrient-rich alluvium, as it slowly dried out, also produced an abundance of fodder for wild herbivores, as well as well as domesticated goats, sheep, and pigs.

The inhabitants of these marshes lived on what are called "turtlebacks," small patches of slightly higher ground, comparable to *cheniers* in the Mississippi delta, often no more than a meter or so above the high-water mark. From these turtlebacks, inhabitants exploited virtually all the wetland resources within reach: reeds and sedges for building and food, a great variety of edible plants (club rush, cattails, water lily, bulrush), tortoises, fish, mollusks, crustaceans, birds, waterfowl, small mammals, and migrating gazelles that provided a major source of protein. The combination of rich alluvial soils with an estuary of two great rivers teeming with nutrients, dead and alive, made for an exceptionally rich riparian life that in turn attracted huge numbers of fish, turtles, birds, and mammals—not to mention humans!—preying on creatures lower on the food chain. In the warm, wet conditions that prevailed in the seventh and sixth millennia BCE, wild subsistence resources were diverse, abundant, stable, and resilient: virtually ideal for a hunter-gatherer-pastoralist.

The density and diversity of resources that are lower in the food chain, in particular, make sedentism more feasible. Compared, say, with hunter-gatherers who may follow large game (seals, bison, caribou), those who take most of their diet

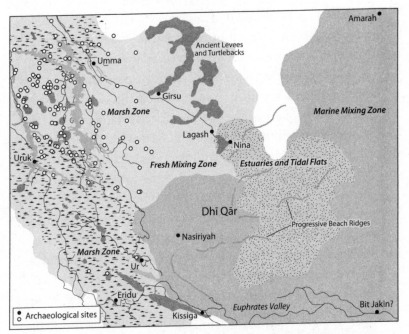

Figure 9. Southern Mesopotamian alluvium:
Ancient watercourses, levees, and turtlebacks, circa 4,500 BCE.
Courtesy of Jennifer Pournelle

from lower trophic levels such as plants, shellfish, fruits, nuts, and small fish that are, other things equal, denser and less mobile than the larger mammals and fish, can be far less migratory. The cornucopia of subsistence resources from lower trophic levels in the wetlands of Mesopotamia was perhaps uniquely favorable to the early creation of substantial sedentary communities.

The first fixed villages in the southern alluvium were not merely in a productive wetland zone; they were located at

the seam of several different ecological zones, allowing villagers to harvest from all of them and to buffer themselves
from the risk of exclusive dependence on any one. They lived
on the border between the water marine environment of the
coast and estuary with its resources and the very different
fresh water ecology of the upstream river environment. The
brackish-water, fresh-water seam, in fact, was a moving border, shifting back and forth with the tides, which, in such flat
terrain, moved great distances. Thus for a large number of
communities, the two ecological zones moved across the landscape while they remained stationary, taking sustenance from
both. The same, even more emphatically, could be said for the
seasons of inundation and drying and the resources particular
to each. A transition between the aquatic resources of the wet
season and the terrestrial resources of the dry season was the
great annual pulse of the region. Instead of the population of
the alluvium having to shift camp from one ecological zone
to another, it could stay in the same place while, as it were,
the different habitats came to them.[14] A subsistence niche in
the southern Mesopotamian wetlands was, compared with the
risks of agriculture, more stable, more resilient, and renewable with little annual labor.

A propitious location and a sense of timing are crucial
to hunter-gatherers in another way. The "harvest" of hunters
and gatherers is less a daily hit-or-miss proposition than a
carefully calculated effort to intercept the roughly predictable
(late-April and May) mass migration of game such as the huge
herds of gazelle and wild asses in the alluvium. The hunt was
carefully prepared in advance. Long, narrowing lanes were
prepared to funnel the herds onto a killing ground, where

they could be dispatched and preserved by drying and salting. For the hunters, as for hunting folk elsewhere, a crucial part of their yearly animal protein supply came from a week or so of intense round-the-clock efforts to take as much migrating prey as practicable. Depending on the setting, the migrating prey in question can comprise large mammals (caribou, gazelle), water fowl (ducks, geese), other migrating birds at their resting or roosting sites, or migrating fish (salmon, eels, alewives, herring, shad, smelt). In many cases the factor limiting the "protein harvest" was not the scarcity of prey but the scarcity of labor to process it before it spoiled. The point is that the rhythm of most hunters is governed by the natural pulse of migrations that represent much of their most prized food supply. Some of these mass migrations of prey may well be a response to human predation, as Herman Melville suggested for the sperm whale, but there is no doubt that it gives a radically different tempo to the lives of hunting and fishing peoples in contrast to agriculturalists—a rhythm that farmers often read as indolence.

The most common route for a great many of these migrations has been via the wetlands, estuaries, and river valleys of major waterways, owing to the density of nutritional resources they offer. Bird migration routes favor marshes and river valleys, as do, more obviously, the movement of anadromous salmon and their mirror image, catadromous eels, to mention only two of the numerous migrating fish species. Any watercourse is itself a nutrient trough with its own flood plains, back swamps, and alluvial fans. The aquatic life along it depends not on its channel but on its periodic invasion of its flood plain (the flood "pulse") for spawning and growth—

making it, in turn, attractive for bird migrations. Thus, for a population located in a rich wetlands at the edge of several ecotones, in a favorable climatic period, and bestriding the intersections of game migration routes for favored prey, its flourishing in the alluvium was perhaps overdetermined. A good many explanations of early sedentism elsewhere have also emphasized the importance of aquatic resources as providing the most favorable conditions for a reliable subsistence.

Exclusive emphasis on the superabundance of marshes and riverine settings overlooks a further crucial advantage of coastal and river locations: transportation. Wetlands may have been a necessary condition of early sedentism, but the development later of large kingdoms and trading centers depended on an advantageous positioning for waterborne trade.[15] The advantage of waterborne transport compared with overland cart or donkey travel is almost impossible to exaggerate. A Diocletian edict specified that the price of a wagon load of wheat doubled after fifty miles. Because it reduces friction dramatically, movement by water is exponentially more efficient.[16] To take the example of firewood, a variety of sources (before railroads and all-weather roads) advise that a cartload of firewood cannot be sold profitably at distance beyond roughly fifteen kilometers—in rugged terrain, even less. The importance of charcoal, though it is massively wasteful of wood, is exclusively due to its superior transportability; its heat value per unit weight and volume is far superior to "raw" firewood. In the premodern era, no bulk goods—timber, metallic ores, salt, grain, reeds, pottery—could be shipped over appreciable distances except by water.

The southern alluvium, in this respect as well, was

uniquely favored. For half the year it was a watery world where transport by reed boats was easy and, being located downstream from the sources of many of the materials the wetland population required, they could take advantage of the current. One must not imagine these early sedentary villages as autarkic economies, consuming only what they produced. Even their hunter-gatherer ancestors were not at all isolated—trading obsidian and prestige goods over substantial distances. The easy availability of waterborne trade in much of the alluvium amplified these exchanges far more than what would have been possible in a landlocked setting.

WHY IGNORED?

Why, one might well ask, were the wetland origins of early sedentary villages and early urbanism overlooked? In part, of course, this is due to the older narrative of civilizations arising from the irrigation of arid lands, a narrative that fit with the contemporary landscape that those formulating the narrative were observing. I believe, however, that the larger context of this historical myopia comes from the nearly indelible association of civilization with the major grains—wheat, barley, rice, maize. (Think of the "amber waves of grain" in "America the Beautiful.") Within this perspective, swamps, marshes, fens, and wetlands generally have been seen as the mirror image of civilization—as a zone of untamed nature, a trackless waste, dangerous to health and safety. The work of civilization, when it came to marshes, was precisely to *drain* them and transform them into orderly, productive grain fields and villages. Civilizing arid lands mean irrigating them; civilizing swamps means draining them; the goal in each case is making arable grain lands. H. R. Hall wrote of early Mesopotamia in "the state of chaos, half-water and half-land, of the [alluvial] fans of southern Babylonia before civilization began its work of draining and canalizing."[17] The work of civilization, or more precisely the state, as we shall see, consists in the elimination of mud and its replacement by its purer constituents, land and water.[18] Whether in ancient China, in the Netherlands, in the fens of England, in the Pontine Marshes finally subdued by Mussolini, or in the remaining southern Iraq marshes drained by Saddam Hussein, the state has endeavored to turn ungovernable wetlands into taxable grain fields by reengineering the landscape.

The absolutely central role of wetland abundance, it merits noting in passing, has not been ignored only in the case of Mesopotamia. Early sedentary communities near Jericho, the earliest settlements in the lower Nile, were wetland-based and only marginally, if at all, dependent on planted grains. Much the same could be said of the Hangzhou Bay, site of the early Neolithic Hemudu culture in the most watery patch of eastern coastal China in the mid-fifth millennium BCE, rich in undomesticated rice—an aquatic plant. Early Indus River settlements, Harrapan and Haripunjaya, fit this description, as do most of the significant Hoabinhian sites in Southeast Asia. Even higher-altitude sites of ancient sedentism—for example, early Teotihuacan near Mexico City or Lake Titicaca in Peru—were set in extensive wetlands that offered abundant harvests of fish, birds, shellfish, and small mammals from the edge environments of several ecologies.

The wetland origins of population settlement have remained relatively invisible for other reasons as well. We are, after all, dealing here with largely oral cultures that left no written records for us to consult. Their relative obscurity is often magnified by the perishable nature of their building materials: reeds, sedges, bamboo, wood, rattan. Even later small societies about which we know from written commentaries by literate neighbors, such as Srivijaya in Sumatra, have been almost impossible to pinpoint, as their remains have been reclaimed by water, soil, and time.

A last and more speculative reason for the obscurity of wetland societies is that they were, and remained, environmentally resistant to centralization and control from above. They were based on what are now called "common property resources"—free-living plants, animals, and aquatic creatures to which the entire community had access. There was no single dominant resource that could be monopolized or controlled from the center, let alone easily taxed. Subsistence in these zones was so diverse, variable, and dependent on such a multitude of tempos as to defy any simple central accounting. Unlike the early states that we will examine later, no central authority could monopolize—and therefore ration—access to arable land, grain, or irrigation water. There was, therefore, little evidence of any hierarchy in such communities (as usually measured by differential grave goods). A culture might well develop in such areas, but the likelihood was small that such an intricate web of relatively egalitarian settlements would throw up great chiefs or kingdoms, let alone dynasties. A state—even a small protostate—requires a subsistence environment that is far simpler than the wetland ecologies we have examined.

MINDING THE GAP

The breathtaking four-millennia gap between the first appearance of domesticated grains and animals and the coalescing of agro-pastoral societies we have associated with early civilization commands our attention. The anomaly of such a stretch of history, when all the building blocks for a classic agrarian society are in place but fail to coalesce, begs an explanation. An implicit assumption of the standard "progress of civilization" narrative is that once domesticated cereals and livestock were available, they would generate, more or less automatically and rapidly, a fully formed agrarian society. As with any new technique, one might anticipate some hesitation as new subsistence routines were accommodated—perhaps even a millennium—but four thousand years, or roughly 160 generations, is far more than a working out of the kinks.

One archaeologist has characterized this long period as one of "low-level food production."[19] Such a term, however, seems singularly inappropriate, as its emphasis on "production" implies a society that is "stuck" at some inferior and unsatisfactory equilibrium. Melinda Zeder, a prominent theorist of domestication, has avoided this teleology in a fashion that implies by contrast that the populations avoiding full reliance on fixed-field cereal crops for the bulk of their caloric needs might actually have known what they were doing: "Stable and highly sustainable subsistence economies based on a mix of free-living, managed, and fully domesticated resources seems to have persisted for 4,000 years or more before the crystallization of agricultural economies based primarily on domestic crops and livestock in the Middle East."[20] In Zeder's view, the Near East was by no means unique in this respect. Citing work on Asia, Meso-America, and eastern North America, she claims that "cultigens and domestic animals were incorporated into the general round of subsistence strategies, sometimes for thousands of years, with little disruption of the traditional hunter-gatherer way of life."

Instead, they served as additional—and often not very important—foods that "differed from wild resources only in that they require propagation rather than hunting or collection to secure them. . . . Thus neither the presence of domesticates or domesticatable resources nor the diffusion of food producing technologies is sufficient to induce the adoption of food production as a guiding principle of subsistence economy."[21]

The first and most prudent assumption about historical actors is that, given their resources and what they know, they are acting reasonably to secure their immediate interests. In this spirit, and because in this case they cannot speak directly for themselves, it makes most sense to see them as agile and astute navigators of a diverse but also changeable and potentially dangerous environment. Just as early sedentism was pioneered by hunters and foragers taking advantage of the multiple subsistence options their diverse wetland setting provided, we can see this long period of as one of continuous experimentation and management of this environment. Rather than relying on only a small bandwidth of food resources, they seem to have been opportunistic generalists with a large portfolio of subsistence options spread across several food webs.

The Mesopotamian alluvium, along with the Levant, is characterized by larger variations in rainfall and vegetation over shorter distances than almost any other place in the world. Seasonal variation in rainfall was also exceptionally high. Although this diversity put different resources fairly close at hand, it also required a large repertoire of subsistence strategies that could be deployed to deal with the variations. There were also the much larger macroclimatic events that, over several millennia, before the first agrarian kingdoms arose around 3,500 BCE, may have made their mark on folk memory of a "great flood." The warmer and wetter period from roughly 12,700 to 10,800 BCE (itself with many oscillations) gave way to the extremely cold (Younger Dryas) period from 10,800 to 9,600 BCE, during which settlements were abandoned and the remaining population retreated to refugia in the warmer bottomlands and on the coasts. Although conditions after the Younger Dryas were generally favorable for hunter-gatherer expansion, there were climatic setbacks such as a century-long period of cold dry weather (beginning around 6,200 BCE) more severe than the Little Ice Age of 1550–1850 known to historians of early modern Europe. Archaeologists of the five or so millennia after 10,000 BCE agree that there were many pulses of population growth and of sedentism: cold and dry periods when sedentism might have been the result of crowding in the available refugia, and warm, wet periods of population growth and dispersion. Given the variation and risks, it would have made no sense for early populations to rely on a narrow bandwidth of subsistence resources.

Thus far we have considered only the climatological and ecological givens and their effect on population distribution and sedentism. It is entirely possible that some or even most of this variation could have had broadly human causes: diseases, epidemics, rapid population growth, exhaustion of local resources and game, social conflict, and violence, not all of which leave unambiguous traces in the archaeological record.

We have surely underestimated the degree of agility and adaptability of our prestate ancestors. This underestimate is built into the civilizational narrative that represents hunter-gatherers, shifting cultivators, and pastoralists virtually as subspecies of Homo sapiens, with each marking a stage of human progress. Yet historical evidence shows that peoples moved fairly readily between these distinctive modes of subsistence and, in fact, combined them in any number of inventive hybrids in the Fertile Crescent and elsewhere. There is evidence, for example, that quasi-sedentary populations in the Mesopotamian alluvium during the Younger Dryas cold spell adopted more mobile subsistence strategies as the abundance of local subsistence forage dwindled, just as, much later, agriculturalists migrating from Taiwan to Southeast Asia (roughly five thousand years ago) often abandoned planting for foraging and hunting in their new and bounteous forest settling.[22] Early in the twentieth century, a major exponent of a geographical perspective on history rejected any categorical distinction among hunter-gatherers, pastoralists, and agriculturalists, emphasizing that for safety's sake, most peoples have preferred to straddle at least two of these subsistence niches — "keeping two strings to their bow in case of necessity."[23]

We should therefore remain militantly agnostic about the basic terms that have animated the historical narratives about the rise of civilizations and of states. Both intellectual skepticism and recent evidence point in this direction. Most discussions of plant domestication and permanent settlement, for example, assume without further ado that early peoples could not wait to settle down in one spot. Such an assumption is an unwarranted reading back from the standard discourses of agrarian states stigmatizing mobile populations as primitive. The "social will to sedentism" should not be taken for granted.[24] Nor should the terms "pastoralist," "agriculturalist," "hunter," or "forager," at least in their essentialist meanings, be taken for granted. They are better understood as defining a spectrum of subsistence activities, not separate peoples, in the ancient Middle East. Kin groups and villages might have pastoralist, hunting, and cereal-growing segments as part of a unified economy. A family or village whose crops had failed might turn wholly or in part to herding; pastoralists who had lost their flocks might turn to planting. Whole areas during a drought or wetter period might radically shift their subsistence strategy. To treat those engaged in these different activities as essentially different peoples inhabiting different life worlds is again to read back the much later stigmatization of pastoralists by agrarian states to an era where it makes no sense. A striking illustration of the shift may be found in Anne Porter's perceptive reading of the many variants of the Epic of Gilgamesh.[25] In the earliest versions, Gilgamesh's soul companion Enkidu is merely a pastoralist, emblematic of a fused society of planters and herders. In versions a millennium later, he is depicted as subhuman, raised among beasts, and requiring sex with a woman to humanize him. Enkidu becomes, in other words, a dangerous barbarian who knows not grain, houses, or cities, or how to "bend the knee." The "late" Enkidu is, as we shall see, the product of the ideology of a mature agrarian state.

Having already domesticated some cereals and legumes, as well as goats and sheep, the people of the Mesopotamian alluvium were already agriculturalists and pastoralists as well as hunter-gatherers. It's just that so long as there were abundant stands of wild foods they could gather and annual migrations of waterfowl and gazelles they could hunt, there was no earthly reason why they would risk relying mainly, let alone exclusively, on labor-intensive farming and livestock rearing. It was precisely the rich mosaic of resources around them and thus their capacity to avoid specializing in any single technique or food source that was the best guarantee of their safety and relative affluence.

WHY PLANT AT ALL?

Yet a good many early Neolithic sites do contain unambiguous evidence of the cultivation of wild cereals and (disputed) evidence of some plant domestication. In the light of the presence in the region of dense wild stands of cereals and other resources, the question becomes not so much why our ancestors didn't plunge headlong into farming, but why they bothered to plant at all. A common answer has been that cereal grains can be harvested, threshed, and stored in a granary for several years and represent a dense store of starches and protein if, by chance, there is a sudden shortage of wild resources. Despite its cost in labor, so the argument goes, it represented something like a subsistence insurance policy for hunter-gatherers who also knew how to plant.

This explanation, in its cruder forms, doesn't hold up to scrutiny. It assumes, implicitly, that the harvest from a planted crop is more reliable than the yield from wild stands of grain. If anything, the opposite is more likely to be the case, inasmuch as wild seed will, by definition, be found only in locations where it will thrive. Second, this perspective overlooks the subsistence risks that the sedentism associated with having to plant, tend, and guard a crop entails. Historically, the subsistence safety of hunters and gatherers lay precisely in their mobility and the diversity of food sources to which they could lay claim. It was, after all, only the rare proximity of so many ecologically varied resources—elsewhere far more temporally and spatially scattered—in the Mesopotamian alluvium that allowed for early sedentism in the first place. If farming further restricted the potential movements of sedentary hunter-gatherers, their inability to respond promptly to, say, an early bird or fish migration may well have diminished rather than enhanced their food security. The periodic evidence throughout this long period of the abandonment of settlement for pastoralism and for migratory foraging attests to sedentism as a strategy rather than the ideology it would later become.

The cruder versions of the "food-storage hypothesis" are also singularly myopic about the great variety of food storage techniques simultaneously practiced in the alluvium and elsewhere.[26] Storage "on the hoof" in the form of livestock is the most obvious. The saying that "the cow is the granary of the Hausa" captures this perfectly. Having a ready supply of fat and protein handy when required may have made small experiments with planting seem less risky and, in fact, some theorists of early agriculture speculate that it was the relative absence of domesticated livestock that helps explain why crop planting spread so much later; it was simply too risky without a reliable fallback. Other foods could also be readily preserved for shorter or longer periods: fish and meat could be salted, dried, and smoked, legumes such as chickpeas and lentils could be dried and stored, fruits and grains could be fermented and distilled. A bowl of fermented barley beer was, apparently, the daily ration for temple laborers in Uruk. From a broader perspective, one might view the landscape as a forager probably saw it: as a massive, diverse, living storage area of fish, mollusks, birds, nuts, fruits, roots, tubers, edible rushes and sedges, amphibians, small mammals, and large game. If one source failed in a given year, another might be abundant. In the diversity and varying temporalities of this living storage complex lay its stability.

One line of theorizing, favored for a time among students of social evolution, depicted agriculture as a crucial civilizational leap because it was a "delayed-return" activity.[27] The cultivator, it asserted, is a qualitatively new person because he must look far ahead in preparing a field for sowing, then must weed and tend the crop as it matures, until (he hopes) it yields a crop. What is wrong—radically so, in my view—is not so much its depiction of the agriculturalist as its caricature of hunter-gatherers. It suggests, by the implied contrast, that the hunter-gatherer is an improvident, spontaneous creature of impulse, coursing the landscape in hope of stumbling on game or finding something good to pluck from a bush or tree ("immediate return"). Nothing could be farther from the truth. All mass capture—gazelle, fish, and bird migrations—involve elaborate, cooperative advance preparation: the building of long narrowing "drive corridors" to a killing ground; building weirs, nets, and traps; building or digging facilities for smoking, drying, or salting of the catch. These are delayed-return activities par excellence. They involve a large kit of tools and techniques and a far greater degree of coordination and cooperation than agriculture requires. Beyond these more spectacular mass-capture activities, hunters and gatherers, as we have seen, have long been sculpting the landscape: encouraging plants that will bear food and raw materials later, burning to create fodder and attract game, weeding natural stands of desirable grains and tubers. Except for the act of harrowing and sowing, they perform all the other operations for wild stands of cereals that farmers do for their crops.

Neither "food storage" nor "delayed return" are remotely plausible reasons for the limited use of domesticated grains that we find in the historical record. I propose a quite different explanation for sowing crops based on a simple analogy between fire and flood. The general problem with farming—especially plough agriculture—is that it involves so much intensive labor. One form of agriculture, however, eliminates most of this labor: "flood-retreat" (also known as décrue or recession) agriculture. In flood-retreat agriculture, seeds are generally broadcast on the fertile silt deposited by an annual riverine flood. The fertile silt in question is, of course, a "transfer by erosion" of upstream nutrients. This form of cultivation was almost certainly the earliest form of agriculture in the Tigris-Euphrates floodplain, not to mention the Nile Valley. It is still widely practiced today and has been shown to be the most labor-saving form of agriculture regardless of the crop being planted.[28]

For our purposes, flooding in this case can be seen to accomplish the same landscape sculpting as the fire deployed by hunter-gatherers or swidden (slash-and-burn) cultivators. A flood clears a "field" by scouring and drowning back all competing vegetation and, in the process, deposits a layer of soft, easily worked, nutritious silt as it recedes. The result, under good conditions, is often a nearly perfectly harrowed and fertilized field ready for sowing at no cost in labor. Just as our ancestors noticed how a fire cleared the land for a new natural succession of quickly colonizing (the so-called r plants) species, so they must have noticed much the same succession with floods.[29] And since the early cereals are grasses (r plants), they would have thrived and gotten a head start on competing weeds if broadcast on this silt. Nor is it much of a stretch, as observed earlier, to imagine making a small breach in a natural levee to provoke a small flood and the recession agriculture that it would make possible. Voila! a form of agriculture that an intelligent, work-shy hunter-gatherer might take up.

CHAPTER TWO

Landscaping the World:
The Domus Complex

ONTRARY to the traditional narrative, there is no magic moment when Homo sapiens crosses some fateful line that separates hunting and foraging from agriculture—from prehistory to history, from savagery to civilization. The moment when a seed or tuber is deposited in prepared soil is more properly seen as one event—and not in itself a very significant one to those doing it—in a long and historically very deep skein of landscape modification starting with Homo erectus and fire.

We, of course, are hardly the only species to modify the environment to our advantage. Although beavers are perhaps the most conspicuous example, elephants, prairie dogs, bears—virtually all mammals, in fact—engage in "niche construction," which changes the physical properties of the landscape and the distribution of other species of flora, fauna, and microbial life around them. Insects, particularly the "social"

insects—ants, termites, bees—do the same. On a broader and deeper historical view, plants are actively engaged in massive landscape modification. Thus the expanding "oak belt" after the last glaciation created, over time, its own soil, shade, fellow-travelling plants, and a supply of acorns that was a boon to dozens of mammals, among them squirrels and Homo sapiens.

Long before what many would consider "proper" agriculture, Homo sapiens had been deliberately rearranging the biotic world around itself with consequences both intended and unintended. Thanks in large part to fire, this low-intensity horticulture practiced over many millennia had a substantial impact on the natural world. As early as eleven or twelve thousand years ago there is firm evidence that populations in the Fertile Crescent were intervening to modify local "wild" plant communities to their advantage many thousands of years before any clear morphological evidence of domesticated grains appears in the archaeological record.[1] We can date the appearance of domesticated grains by the telltale complex of weedy species characteristic of active tilling and tending of cultivated fields that appears simultaneously, as does the apparent decline of indigenous flora less adapted to this managed environment.[2]

Nowhere has evidence of landscape sculpting had more impact than in our understanding of the early peopling of the forests of the Amazonian floodplain. There, it now appears that the basin was well populated and made habitable in large part owing to landscape management of palms, fruit trees, Brazil nuts, and bamboos that gradually created culturally anthropogenic forests. Given sufficient time to work its magic, slow motion forest "gardening" of this sort can create the soils, flora, and fauna that represent an abundant subsistence niche.[3]

Planting a seed or tuber is, in this context, only one of hundreds of techniques designed to increase the productivity, density, and health of desirable but morphologically wild plants. Some of these techniques include the burning of undesirable flora, weeding wild stands of favored plants and trees to eliminate competitors, pruning, thinning, selective harvesting, trimming, transplanting, mulching, relocating protective insects, bark-ringing, coppicing, watering, and fertilizing.[4] For animals, short of full domestication, hunters have long been burning to encourage browse for prey, sparing females of reproductive age, culling, hunting based on life cycles and population, fishing selectively, managing streams and other waters to promote spawning and shellfish beds, transplanting the eggs and young of birds and fish, manipulating habitat, and occasionally raising juveniles.

Domestication, in light of the deep history and massive effects of these practices, needs to be seen far more expansively than mere planting and pastoralism. Since the dawn of the species, Homo sapiens has been domesticating whole environments, not just species. The preeminent tool for this, before the Industrial Revolution, was not the plough so much as fire. The domestication of whole environments in turn made possible the other adaptive advantage of our species, namely high rates of reproduction, making us the world's most successful invasive mammal (of which more later). Whether we wish to call it niche construction, domestication of the environment, landscape modification, or the human management of ecosystems, it is clear on a long view that much of the world was shaped by human activity (anthropogenic) well before the first societies based on fully domesticated wheat, barley, goats, and sheep appear in Mesopotamia. This is why, finally, the conventional "subspecies" of subsistence modes—hunting, foraging, pastoralism, and farming—make so little historical sense. The same people have practiced all four, sometimes in a single lifetime; the activities can and have been combined for thousands of years, and each of them bleeds imperceptibly into the next along a vast continuum of human rearrangements of the natural world.

FROM NEOLITHIC PLANTING TO FLORAL ZOO: CONSEQUENCES OF CULTIVATION

Even if the search for a decisive moment in the domestication of the earliest grains is a pointless endeavor, there is no doubt at all that by 5,000 BCE there were hundreds of villages in the Fertile Crescent cultivating fully domesticated grains as their main staple. Why this should be so is a puzzle around which dispute still swirls. The dominant explanation until fairly recently was what might be called the "backs-to-the-wall" theory of plough agriculture associated with the great Danish economist Ester Boserup.[5] Starting from the unassailable premise that plough cultivation typically required far more work for the calories it returned than did hunting and gathering, she reasoned that full cultivation was taken up not as an opportunity but as a last resort when no other alternative was possible. Some combination of population growth, the decline in wild protein to hunt and nutritious wild flora to gather, or coercion, must have forced people, reluctantly, to work harder to extract more calories from the land they had access to. This demographic transition to drudgery has been read by many as metaphorically captured in the biblical tale of Adam and Eve being expelled from Eden to a world of toil.

Despite its apparent economic logic, the backs-to-the-wall thesis, at least in Mesopotamia and the Fertile Crescent, fails to match the available evidence. One would expect cultivation to be adopted first in those areas where hard-pressed foragers had reached the carrying capacity of their immediate environment. Instead, it seems to have arisen in areas characterized more by abundance than by scarcity. If, as noted earlier, they were practicing flood-retreat agriculture, then the central premise of the Boserupian argument of cultivation requiring great toil may well be invalid. Finally, there appears to be no firm evidence associating early cultivation with the disappearance of either game animals or forage. The backs-to-the-wall theory of agriculture is in tatters (at least for the Middle East), but it has not been replaced by a satisfactory alternative explanation for the spread of cultivation.[6]

THE DOMUS AS A MODULE OF EVOLUTION

The question itself may be less important than supposed. So long as it was not terribly labor intensive, cultivation may have been one of many techniques of environmental engineering in early sedentary communities. What seems more important than why sowing and tilling crops became more common are the far-reaching consequences of grain and animal domestication once accomplished: a subject to which we now turn.

Whatever the reasons for the growing reliance on domesticated grains and animals for subsistence, it represented a qualitative change in landscape modification. The cultivars were transformed; the livestock was transformed; the soils and fodder they depended on were transformed; and, not least, Homo sapiens was transformed. Here the term "domestication"—from "domus," or household—needs to be taken rather literally. The domus was a unique and unprecedented concentration of tilled fields, seed and grain stores, people, and domestic animals, all coevolving with consequences no one could have possibly foreseen. Just as important, the domus as a module of evolution was irresistibly attractive to literally thousands of uninvited hangers-on who thrived in its little ecosystem. At the top of the heap were the so-called commensals: sparrows, mice, rats, crows, and (quasi-invited) dogs, pigs, and cats for which this new Ark was a veritable feedlot. Each of these commensals in turn brought along its own train of microparasites—fleas, ticks, leeches, mosquitoes, lice, and mites—as well as their predators; the dogs and cats were there in large part for the mice, rats, and sparrows. Not a single critter emerged from its sojourn at the late-Neolithic multispecies resettlement camp unaffected.

Archaeo-botanists have devoted most attention to the morphological and genetic changes in the major grains: wheat and barley. The early wheats—einkorn and, especially, emmer—along with barley and most of the "founder" pulses—lentils, peas, chickpeas, bitter vetch, and even flax—could be said to belong broadly to the "grain" family, as they are self-pollinating annuals and do not readily cross with their wild progenitors (unlike rye). Many plants are quite finicky about where and when they will grow. Those most eligible for domestication were, aside from their food value, "generalists" that could thrive in disturbed soils (the tilled field), could grow in dense stands, and were easily stored. The problem for the would-be farmer was that the natural selection pressure for wild plants promotes characteristics that are designed to defeat the farmer. Thus wild grainheads are typically small and shatter easily, thereby seeding themselves. They mature unevenly; their seeds can remain long dormant but still germinate; they have many appendages, awns, glumes, and thick seed coats, all of which discourage grazers and birds. All these features are selected for in the wild and selected against by the farmer. It is diagnostic that the major weeds that plague wheat and barley—one can think of them as hitchhiking, feral commensals—have precisely these characteristics. They like the tilled field but escape the harvester and grazer alike. Oats apparently began their agricultural career as a weed (an obligate pest mimicking the crop) in the tilled field and eventually became a secondary crop.

The tilled, sown, weeded field is an altogether different terrain of selection. The farmer wants nonshattering (indehiscent) grain spikes that can be gathered intact, as well as determinate growth and maturity. Many of the characteristics of a domestic grain are simply the long-run effects of sowing and harvesting. Thus plants that produce both more seeds and seeds that are larger, with thin coats (allowing them to quickly germinate and outrace weedy competitors when sown), that ripen uniformly, are easily threshed, germinate reliably, and have fewer glumes and appendages are likely to contribute disproportionately to the harvest, and thus their offspring will be favored in next year's planting. The morphological differences between the continuously selected, planted cultivar and its wild progenitor become massive over time. In wheats, the difference between wild and domesticated varieties is easily apparent but not as striking as the contrast between maize and its primitive ancestor, teosinte, which it is hard to imagine belongs to the same species at all.

The early agricultural field was vastly more simplified and "cultivated" than the world outside it. At the same time it was far more complex than industrial field agriculture, with its sterile hybrids and clones grown largely for yield. Early agriculture was something of a portfolio of cultivars and land races that were grown for more than one purpose and were deliberately chosen not so much for their average yield as for their resistance to various stresses, diseases, and parasites and their reliability in meeting subsistence needs. The diversity of crops and subspecies was greatest in natural settings of greater ecological and climatic diversity and least in the alluvial bottomlands with more dependable water and growing conditions.

The purpose of the cultivated field and of the garden is precisely to eliminate most of the variables that would compete against the cultigen. In this man-made and -defended environment—other flora, exterminated for a time by fire, flood, plough, and hoe, pulled out by their roots; birds, rodents, and browsers scared off or fenced out—we make a nearly ideal world in which our favorites, perhaps carefully watered and fertilized, will flourish. Steadily, by coddling, we create a fully domesticated plant. "Fully domesticated" means simply that it is, in effect, our creation; it can no longer thrive without our attentions. In evolutionary terms a fully domesticated plant has become a superspecialized floral "basket case," and its future is entirely dependent on our own. If it ceases to please us, it will be banished and almost certainly will perish.[7] Some domestic plants and animals (oats, bananas, daffodils, day lilies, dogs, and pigs) have, as we know, resisted full domestication and are capable, to varying degrees, of surviving and reproducing outside the domus.

FROM HUNTER'S PREY
TO FARMER'S CORRAL

We can surely understand how dogs, cats, and even pigs have been attracted to hunters and to the domus for the food, warmth, and concentration of available prey they promised. They—some of them at any rate—appeared at the domus more as volunteers than as conscripts. Much the same could be said for the house mouse and the house sparrow, which, though perhaps less welcome, came while evading full domestication. The case of the sheep and goat, the first noncommensal domesticates in the Middle East, however, constitutes a profound revolution in mammalian affairs. Here were, after all, animals that for many thousands of years were the prey of Homo sapiens the hunter. Instead of merely killing them, Neolithic villagers captured them, penned them, protected them from other predators, fed them when necessary, bred them to increase their progeny, used the milk, wool, and blood of the living animal and then used the carcass of the slaughtered animal as a hunter might. The transition from prey to "protected" or "cultivated" species was freighted with enormous consequences for both parties to the transaction.

If Homo sapiens is judged the most successful and numerous invasive species in history, this dubious achievement has been due to the allied battalions of domesticated plants and livestock it has taken with it to virtually every corner of the globe.

Not all prey animals were suitable candidates. Here the evolutionary biologists and natural historians stress that certain species were "preadapted," having characteristics in the wild that predisposed them to life in the domus. Among the characteristics proposed are, above all, herd behavior and the social hierarchy that accompanies it,[8] the capacity to tolerate different environmental conditions, a broad spectrum diet, adaptability to crowding and disease, the ability to breed under confinement, and, finally, a relatively muted fright-and-flight response to external stimuli. While it is true that most major domesticates (sheep, goat, cattle, and pigs) are herd animals, as are most domesticated draft animals (horses, camels, donkeys, water buffalo, and reindeer), herd behavior does not guarantee domestication. The gazelle, for example, was by far the most frequently hunted animal for several millennia. Long, guiding, funnel-shaped walls (called desert kites) are found in northern Mesopotamia, designed to intercept their annual migratory herds. Unlike the sheep, goats, and cattle, however, this source of desirable protein does not survive under domestication.

Those animals that were domesticated, however, entered an entirely new life world, encountering radically different evolutionary pressures from those they had experienced as free-living prey. First and foremost, to take the most common early domesticates, sheep, goats, and pigs, they were not free to go wherever they pleased. As a captive species their diet was, along with their mobility, restricted, and they were often crowded together in enclosures, wadis, and caves to a degree unprecedented in their evolutionary history. Crowding had, as we shall see, consequences for their health and social organization. One major goal of their captors was to maximize their reproduction. This was typically achieved, as it is in the modern flock, by culling both young males and females beyond reproductive age in order to maximize the number of fertile females and their progeny. When archaeologists wish to know whether a large find of sheep or goat bones is from a wild or domesticated flock, the age and gender distribution of the remains provides the strongest evidence of active human management and selection. While guarded and tended by their human masters, the domesticates, like plants in the field, were spared many of the selective pressures (predators, competition for food, battles for mates) of the wild but were subject to new selection pressure, both deliberate and unintentional, imposed by their "owners."[9]

The new terrain of selection cannot be confined to the designs of Homo sapiens but applies more broadly to the microecology and microclimate of the entire domus complex: its fields, its crops, its shelters, and the massive cavalcade of animals, birds, insects, and parasites down to bacterial life that were assembled there as commensals. Proof of the independent effect of the domus complex, independent of direct human management, is that uninvited commensals such as mice, sparrows, and even pigs (who might have also come on their own to forage in the rich pickings of human settlement) exhibit some of the same physical changes as full domesticates.[10]

Subject to radical new pressures at the domus, the major domesticates became different animals, both physiologically and behaviorally. These changes, furthermore, occurred in what was, in evolutionary terms, the blink of an eye. We know this in part by comparing skeletal remains of domesticated animals in Mesopotamia with the remains of their wild cousins and progenitors, as well as by more contemporary experiments in domestication. The now famous Russian experiment in the taming of silver foxes is a striking example. By selecting the least aggressive (most tame) from among 130 silver foxes and breeding them to one another repeatedly, the experimenters produced, in only ten generations, 18 percent of progeny that exhibited extremely tame behavior—whining, wagging their tails, and responding favorably to petting and handling as a domestic dog might. After twenty generations of such breeding, the percentage of extremely tame foxes nearly doubled to 35 percent.[11] The behavioral transformation was accompanied by physical changes such as lop ears, piebaldness, and a raised tail that some see as linked genetically to the decrease in adrenaline production.

The hallmark behavioral difference between domesticated animals and their wild contemporaries is a lower threshold of reaction to external stimuli and an overall reduced wariness of other species—including Homo sapiens.[12] The likelihood that such traits are in part a "domus effect" rather than entirely due to conscious human selection is, once again, suggested by the fact that uninvited commensals such as statuary pigeons, rats, mice, and sparrows exhibit much the same reduced wariness and reactivity. Selection, for example, favored smaller, less obtrusive rats and mice better adapted to living off human refuse and avoiding detection and capture. As a sheep breeder myself for more than twenty years, I have always been personally offended when sheep are used as a synonym for cowardly crowd behavior and a lack of individuality. We have, for the past eight thousand years, been selecting among sheep for tractability—slaughtering first the aggressive ones who broke out of the corral. How dare we, then, turn around and slander a species for some combination of normal herd behavior and precisely those characteristics we have selected for?

Associated with this process of behavioral change are a variety of physical changes. They typically include a reduction in male-female differences (sexual dimorphism). Male sheep horns, for example, diminish or disappear altogether because they are no longer selected to ward off predators or to compete for breeding mates. Domesticates are far more fertile than their wild cousins. Another common and striking morphological change among domesticates is known as neotany: the relatively early attainment of adulthood by many domesticates and their retention, as adults, of much of the juvenile morphology—especially the skull—and juvenile behaviors of their free-living ancestors. A shortening of the face and jaw results in shorter molars and, as it were, a more crowded skull.

The reduction in brain size and, somewhat more speculatively, its consequences, seem decisive for the ensemble of what we might call "tameness" among domestic animals generally. Compared with their wild ancestors, sheep have undergone a reduction in brain size of 24 percent over the ten thousand-year history of their domestication; ferrets (domesticated far more recently) have brains 30 percent smaller than those of wild polecats; and pigs (*sus scrofa*) have brains more than a third smaller than their ancestors'.[13] At the new frontier of domestication—aquaculture—even captive-reared rainbow trout have smaller brains than do wild trout.

More diagnostic than the overall reduction in brain size are the areas of the brain that seem to be disproportionately affected. In the case of dogs, sheep, and pigs, the part of the brain most affected is the limbic system (hippocampus, hypothalamus, pituitary, and amygdala), which is responsible for activating hormones and nervous-system reactions to threats and external stimuli. The shrinkage of the limbic system is associated with raising the threshold that would trigger aggression, flight, and fear. In turn, this helps explain the diagnostic characteristics of virtually all domesticated species: namely the general reduction in emotional reactivity. Such emotional dampening can be seen as a condition for life in the crowded domus and under human supervision, where the instant reaction to predator and prey are no longer powerful pressures of natural selection. With physical protection and nutrition more secure, the domesticated animal can be less intently alert to its immediate surroundings than its cousins in the wild.

Just as human sedentism represents a reduction in mobility and increased crowding in the village and domus, so the relative confinement and crowding of domestic animals has immediate consequences for health. The stress and physical trauma of confinement, together with a narrower spectrum diet and the ease with which infections can spread among individuals of the same species packed together, make for a variety of pathologies. Bone pathologies due to repeated infection, relative inactivity, and a poorer diet are particularly common. Archaeologists have come to expect cases of chronic arthritis, evidence of gum disease, and bone signatures of confinement in analyzing the remains of archaic domestic animals. The result is also far higher mortality rates among newborn domesticates. Among confined llamas, for example, the mortality rate for newborns approaches 50 percent, far higher than among wild llamas (guanacos). The difference can be largely attributed to the effects of confinement—muddy, feces-rich corrals in which virulent clostridium bacteria, among others, thrives and, like other parasites, finds an abundant supply of hosts close at hand.

The high rates of mortality for newborn domesticates would seem to defeat the purpose of human management, which is largely to maximize the reproduction of animal protein as one maximizes one's crop of grain. It appears, however, that the rates of fertility may increase so dramatically as to more than offset the losses through mortality. The reasons are not entirely clear, but domesticated animals generally reach reproductive age earlier, ovulate and conceive more frequently, and have longer reproductive lives. Tame silver foxes in the Russian experiment came into heat twice a year compared with once a year for undomesticated foxes. The pattern for rats is more striking, although as commensals even in their wild state, they allow only speculative inferences to other domesticates. Captured wild rats have quite low rates of fertility, but after only eight (short!) generations of captivity, their rate of fertility was found to increase from 64 percent to 94 percent and by the twenty-fifth generation, the reproductive life of captive rats was twice as long as "noncaptives."[14] They were, overall, nearly three times as fecund. The paradox of relative ill health and high newborn mortality on the one hand, coupled with more-than-compensating increases in fertility on the other, is one to which we shall return, as it bears directly on the demographic explosion of agricultural peoples at the expense of hunters and gatherers.

SPECULATION ON HUMAN PARALLELS

To what degree is it plausible to look for analogous changes in morphology and behavior as Homo sapiens adapted to sedentism, crowding, and an increasingly cereal-dominated diet? This path of inquiry is as speculative as it is intriguing. But it is, I believe, fruitful precisely because it entertains the idea that we are as much a product of self-domestication in both intended and unintended ways as other species of the domus are products of our domestication.

One way of determining whether a woman who died nine thousand years ago was living in a sedentary, grain-growing community as compared with a foraging band was simply to examine the bones of her back, toes, and knees. Women in grain villages had characteristic bent-under toes and deformed knees that came from long hours kneeling and rocking back and forth grinding grain. It was a small but telling way that new subsistence routines—what today would be called a repetitive stress injury—shaped our bodies to new purposes, much as the work animals domesticated later—cattle, horses, and donkeys—bore skeletal signature of their work routines.[15]

The analogies are potentially far-reaching. One might argue that the spread of sedentism transformed Homo sapiens into far more of a herd animal than previously. Unprecedented concentrations of people, as in other herds, provided ideal conditions for epidemics and the sharing of parasites. But this aggregation was not a one-species herd but an aggregation of many mammalian herds who shared pathogens and generated entirely new zoonotic diseases by the mere fact of being assembled around the domus for the first time. Hence the term "late-Neolithic multispecies resettlement camp." We were all, one might say, crowded onto the same ark, sharing its microenvironment, sharing our germs and parasites, breathing its air.

No wonder then that the archaeological signs for a life lived largely in the domus are strikingly similar for man and beast. "Domiciled" sheep, for example, are generally smaller than their wild ancestors; they bear telltale signs of domesticate life: bone pathologies typical of crowding and a narrow diet with distinctive deficiencies. The bones of "domiciled" Homo sapiens compared with those of hunter-gatherers are also distinctive: they are smaller; the bones and teeth often bear the signature of nutritional distress, in particular, an iron-deficiency anemia marked above all in women of reproductive age whose diets consist increasingly of grains.

The parallel, of course, arises from a common environment of more restricted mobility, crowding and the cross-infection opportunities it presents, a narrower diet (less variety for herbivores, less variety and less protein for omnivores like Homo sapiens), and relaxation of some of the selection pressures from predators lurking outside the domus. In the case of Homo sapiens, however, the process of self-domestication had begun long before (some of it even before "sapiens") with the use of fire, cooking, and the domestication of grain. Thus declining tooth size, facial shortening, a reduc-

tion in stature and skeletal robustness and less sexual dimorphism were evolutionary effects that had a far longer history
than the Neolithic alone. Nevertheless, sedentism, crowding,
and a diet increasingly dominated by cereals were revolutionary changes that left an immediate and legible mark on the
archaeological record.

The possibility that domestication in the largest sense is
an analogous process that we can see at work among humans
and their domesticates has been put most forcefully and eloquently by Helen Leach.[16] She notes the similar trends since
the Pleistocene in size, stature (grain diets are typically associated with shorter stature), tooth-size reduction, and shortening of face and jaws and asks pointedly whether there might
be a "distinctive syndrome" of domestication arising from the
increasingly common environment that they share. By "common environment" she means not merely sedentism and grain
but the entire assemblage of the domus. We might think of it
as a "domus module," one that would eventually go on to colonize much of the world.[17]

By viewing domestication in its broadest sense as accli-
matization to life in a household, and extending that concept
to incorporate the house and the outbuildings, yards, gardens
and orchards, we can consider some of the criteria of domestication as biological changes brought about through living in
the culturally modified, artificial environment which we call
the domus.

> The complex of houses and yards protected all of the settle-
> ment's inhabitants in the winter months, including invited and
> uninvited commensals. Tidbits, scraps, or spoiled items, foods
> prepared from pounded and ground plant parts reached the

dogs and, later in the Neolithic era the pigs kept in the household compounds. A shared diet between humans, dogs, and
pigs—one that was becoming softer in consistency—might
partly explain shared gracilization [loss of bone mass due to
evolution] and cranio-facial and dental reduction in these
species.[18]

Beyond the morphological and physiological consequences of domestication for man and beast lie changes in
behavior and sensibility that are more difficult to codify. The
physical and cultural realms are closely connected. Is it the
case, for example, that like their domesticates, sedentary,
grain-planting, domus-sheltered people have experienced a
comparable decline in emotional reactivity and are less intently alert to their immediate surroundings? If so, is it related, as in domestic animals, to changes in the limbic system,
which governs fear, aggression, and flight responses? I know
of no evidence bearing directly on this question, nor is it easy
to imagine how the question could be addressed in an objective way.

As far as biological changes associated with agriculture
itself are concerned, we must be doubly cautious. Selection
works by variation and inheritance, and only 240 human generations have elapsed since the first adoption of agriculture
and perhaps no more than 160 generations since it became
widespread. We are, therefore, hardly in a position to reach
sweeping conclusions.[19] While issues of this scope may be beyond our capacity to resolve, we may be able to say more about
how sedentism, animal and plant domestication, and a largely
grain diet has shaped our behavior, routines and our health.

THE DOMESTICATION OF US

We, as a species, are inclined to see ourselves as the "agent" in narratives of domestication. "We" domesticated wheat, rice, the sheep, the pig, the goat. But if we squint at the matter from a slightly different angle, one could argue that it is we who have been domesticated. Michael Pollan sees it this way in his sudden and memorable aperçu while gardening.[20] As he is weeding and hoeing around his thriving potato plants, it dawns on him that he has, unwittingly, become the slave of the potato. Here he is, on his hands and knees, day after day, weeding, fertilizing, untangling, protecting, and in general reshaping the immediate environment to the utopian expectations of his potato plants. Looked at from this angle, who is doing whose bidding becomes almost a problem in metaphysics. If our domesticated plants cannot thrive without our help, it is equally true that our survival as a species has likewise become dependent on a handful of domesticated cultivars.

The domestication of animals can be seen in virtually identical terms. Who is serving whom is no simple matter while cattle and other livestock are being reared, led to pasture, given fodder, and protected. Evans-Pritchard, in his famous monograph on the ultimate cattle people, the Nuer, had much the same insight about the Nuer and their cattle as Pollan had about his potatoes.

> It has been remarked that the Nuer might be called parasites of the cow. But it might be said with equal force that the cow is a parasite of the Nuer, who lives are spent in insuring its welfare: they build byres, kindle fires, and clear kraals for its comfort,

move from villages to camps, from camp to camp, from camps back to villages for its health, defy wild beasts for its protection and fashion ornaments for its adornment. It lives its gentle, indolent, sluggish life thanks to the Nuer's devotion.[21]

One might well object to this line of reasoning by observing that, in the final analysis, Pollan eats his potato and the Nuer eat (trade, barter, and tan the skin of) their cattle. The final disposition is not in doubt. But this overlooks the fact that while it lives, the potato and the cow are the objects of a demanding and solicitous routine that caters to their well-being and safety.

Thus, while larger questions of how our brains and limbic systems have been shaped by domestication cannot yet be determined, we can nevertheless say something about how life in the late Neolithic has been shaped by our relationship to our domesticates in the domus.

First let us compare, broadly, the life world of the hunter-forager with that of the farmer, with or without livestock. Close observers of hunter-gatherer life have been struck by how it is punctuated by bursts of intense activity over short periods of time. The activity itself is enormously varied—hunting and collecting, fishing, picking, making traps and weirs—and designed in one way or another to take best advantage of the natural tempo of food availability. "Tempo," I think, is the key word here. The lives of hunter-gatherers are orchestrated by a host of natural rhythms of which they must be keen observers: the movement of herds of game (deer, gazelle, antelope, pigs); the seasonal migrations of birds, especially waterfowl, which can be intercepted and netted at their resting or nesting places; the runs of desirable fish upstream or downstream; the cycles of the ripening of fruits and nuts, which must be collected before other competitors arrive or before they spoil; and, less predictably, appearances of game, fish, turtles, and mushrooms, which must be exploited quickly. The list could be expanded almost indefinitely, but several aspects of this activity stand out. First, each activity requires a different "tool kit" and techniques of capture or collecting that must be mastered. Second, we should not forget that foragers have long gathered grains from natural stands of cereals and had, for this purpose, already developed virtually all the tools we associate with the Neolithic tool kit: sickles, threshing mats and baskets, winnowing trays, pounding mortars and grinding stones, and the like. Third, each of these activities represents a distinct problem in coordination such that the cooperative group and division of labor for each is different. Finally, the activities, like those of the earliest village in the Mesopotamian alluvium, span several food webs—wetlands, forest, savanna, and arid—each of which has its own distinct seasonality. While hunter-gatherers depend vitally on these rhythms, they are, at the same time, generalists and opportunists ever alert to take advantage of the scattered and episodic bounty nature may bring their way.

Botanists and naturalists have been continually amazed by the degree and breadth of knowledge hunters-gatherers have of the natural world around them. Their taxonomies of plants are not classified in Linnaean categories, but they are both more practical (good to eat, will heal wounds, will make blue dye) and quite as elaborate.[22] Codifications of farming knowledge in America, by contrast, have traditionally taken the form of the *Farmers' Almanac*, which suggests, among other things, when maize should be planted. We might, in this context, think of hunters and gatherers as having an entire library of almanacs: one for natural stands of cereals, subdivided into wheats, barleys, and oats; one for forest nuts and fruits, subdivided into acorns, beechnuts, and various berries; one for fishing, subdivided by shellfish, eels, herring, and shad; and so on. What is perhaps just as astonishing is that this veritable encyclopedia of knowledge, including its historical depth of past experience, is preserved entirely in the collective memory and oral tradition of the band.

To return to the concept of tempo, one might think of hunters and gatherers as attentive to the distinct metronome of a great diversity of natural rhythms. Farmers, especially fixed-field, cereal-grain farmers, are largely confined to a single food web, and their routines are geared to its particular tempo. Bringing a handful of crops successfully to harvest is to be sure a demanding and complex activity, but it is usually dominated by the requirements of one dominant starch plant. It is no exaggeration to say that hunting and foraging are, in terms of complexity, as different from cereal-grain farming as cereal-grain farming is, in turn, removed from repetitive work on a modern assembly line. Each step represents a substantial narrowing of focus and a simplification of tasks.[23]

The domestication of plants as represented ultimately by fixed-field farming, then, enmeshed us in an annual set of routines that organized our work life, our settlement patterns, our social structure, the built environment of the domus, and much of our ritual life. From field clearing (by fire, plough, harrow), to sowing, to weeding, to watering, to constant vigilance as the crop ripens, the dominant cultivar organizes much of our timetable. The harvest itself sets in train another sequence of routines: in the case of cereal crops, cutting, bundling, threshing, gleaning, separation of straw, winnowing chaff, sieving, drying, sorting—most of which has historically been coded as women's work. Then, the daily preparation of grains for consumption—pounding, grinding, fire making, cooking, and baking throughout the year—set the tempo of the domus.

These meticulous, demanding, interlocked, and mandatory annual and daily routines, I would argue, belong at the center of any comprehensive account of the "civilizing process." They strap agriculturalists to a minutely choreographed routine of dance steps; they shape their physical bodies, they shape the architecture and layout of the domus; they insist, as it were, on a certain pattern of cooperation and coordination. In that sense, to pursue the metaphor, they are the background musical beat of the domus. Once Homo sapiens took that fateful step into agriculture, our species entered an austere monastery whose taskmaster consists mostly of the demanding genetic clockwork of a few plants and, in Mesopotamia particularly, wheat or barley.

Norbert Elias wrote convincingly of the growing chains of dependence among ever denser populations in medieval Europe that made for the mutual accommodation and restraint that he termed "the civilizing process."[24] But literally thousands of years before the social changes Elias describes—and quite apart from any hypothetical changes to our limbic system—much of our species was already disciplined and subordinated to the metronome of our own crops.

Once cereals became established as a staple in the early

Middle East, it is striking how the agricultural calendar came to determine much of public ritual life: ceremonial ploughing by priests and kings, harvest rites and celebrations, prayers and sacrifices for an abundant harvest, gods for particular grains. The metaphors with which people reasoned were increasingly dominated by domesticated grains and domesticated animals: "a time to sow and a time to reap," being "a good shepherd." There is hardly a passage in the Old Testament that fails to make use of such imagery. This codification of subsistence and ritual life around the domus was powerful evidence that, with domestication, Homo sapiens had traded a wide spectrum of wild flora for a handful of cereals and a wide spectrum of wild fauna for a handful of livestock.

I am tempted to see the late Neolithic revolution, for all its contributions to large-scale societies, as something of a deskilling. Adam Smith's iconic example of the productivity gains achievable through the division of labor was the pin factory, where each minute step of pin making was broken down into a task carried out by a different worker. Alexis de Tocqueville read *The Wealth of Nations* sympathetically but asked, "What can be expected of a man who has spent twenty years of his life putting heads on pins."[25]

If this is a too bleak view of a breakthrough credited with making civilization possible, let us at least say that it represented a contraction of our species' attention to and practical knowledge of the natural world, a contraction of diet, a contraction of space, and perhaps a contraction, as well, in the breadth of ritual life.

CHAPTER THREE

Zoonoses: A Perfect
Epidemiological Storm

DRUDGERY AND ITS HISTORY

AGRO-PASTORALISM—ploughed fields and domestic animals—comes to dominate much of Mesopotamia and the Fertile Crescent well before the appearance of states. With the exception of areas favored by flood-retreat agriculture, this fact represents a paradox that, in my view, has still not been satisfactorily explained. Why would foragers in their right mind choose the huge increase in drudgery entailed by fixed-field agriculture and animal husbandry unless they had, as it were, a pistol at their collective temple? We know that even contemporary hunter-gatherers, reduced to living in resource-poor environments, still spend only half their time in anything we might call subsistence labor. As the students of a rare archaeological site in Mesopotamia (Abu Hureyra), where the entire transition from hunting and gathering to full-blown agriculture can be traced, put it, "No hunter-gatherers occupying a productive locality with a range of wild foods able to provide for all seasons are likely to have started cultivating their caloric staples willingly. Energy investment per unit of energy return would have been too high."[1] Their conclusion was that the "pistol at their temple" in this case was the cold snap of the Younger Dryas (10,500–9,600 BCE), which reduced the abundance of wild plants, together with hostile adjacent populations, which restricted their mobility. This explanation, as noted earlier, is hotly contested in terms of both evidence and logic.

I am in no position to adjudicate, let alone resolve, the controversy over what drove people over several millennia to agriculture as a dominant mode of subsistence. The long-accepted explanation, virtually an orthodoxy, was an intellectually satisfying narrative of subsistence intensification covering a span of as much as six thousand years. The first pulse of intensification was termed "the broad spectrum revolution," a reference to the exploitation of more varied subsistence resources at lower trophic levels. The transition was brought about in the Fertile Crescent by the growing scarcity (by overhunting?) of the big-game sources of wild protein—aurochs, onager, red deer, sea turtle, gazelle—the "low-hanging fruit," to mix metaphors, of early hunting. The result, perhaps impelled as well by population pressure, forced people to exploit resources that, while abundant, required more labor and were perhaps less desirable and/or nutritious. Evidence for this broad-spectrum revolution is ubiquitous in the archaeological record as the bones of large wild animals decline and the volume of starchier plant matter, shellfish, small birds and mammals, snails, and mussels begin to predominate. For the founders of this orthodoxy, the logic behind the broad-spectrum revolution and the adoption of agriculture was identical and, moreover, worldwide. The global increase in population, especially after 9,600 BCE, when the climate improved, together with the decline in big game (clearly documented in the Middle East and the New World), forced hunters and gatherers to intensify their foraging. Pressing ever more heavily on the carrying capacity of their environment's resources, they were obliged to work harder for their subsistence. Thus the broad-spectrum revolution was, in this view, the first step in a long increase in drudgery that later reached its logical conclusion in the even more unremitting toil of plough agriculture and livestock rearing. In most versions of this narrative, the broad-spectrum revolution and agriculture were also nutritionally damaging, resulting in poorer health and higher mortality.

As an explanation for the broad-spectrum revolution, demographic pressure on carrying capacity seems in many locations to be in conflict with the available evidence. The "revolution" occurs in settings where there seems to be little population pressure on resources. It may also be the case that the wetter and warmer conditions after 9,600 BCE promoted a much greater abundance of plant life, as in the Mesopotamian alluvium, that could be easily gathered, though this would not explain the observed nutritional deficiencies in the archaeological record. There is no doubting the reality of the broad-spectrum revolution, but the jury is still out when it comes to understanding either its causes or its consequences.

About the development of agriculture proper, some three or four millennia later, however, the jury is in. There was growing population pressure; sedentary hunters and gatherers found it harder to move and were impelled to extract more, at a higher cost in labor, from their surroundings, and most large game was in decline or gone. This, then, is no Whiggish

story of human invention and progress. Planting techniques
were long known and occasionally used; wild plants were rou-
tinely gathered and their seeds stored; all the tools for grain
processing were at hand, and even a captive animal or two
might be held in reserve. Nevertheless, planting and livestock
rearing as *dominant* subsistence practices were avoided for as
long as possible because of the work they required. And most
of the work arose from the need to defend a simplified, artificial landscape from the resurgence of nature excluded from it:
other plants (weeds), birds, grazing animals, rodents, insects,
and the rust and fungal infections that threatened a mono-
cropped field. The tilled agricultural field was not only labor
intensive; it was fragile and vulnerable.

THE LATE NEOLITHIC MULTISPECIES
RESETTLEMENT CAMP: A PERFECT
EPIDEMIOLOGICAL STORM

The world's population in 10,000 BCE, according to one
careful estimate, was roughly 4 million. A full five thousand
years later, in 5,000 BCE, it had risen only to 5 million. This
hardly represents a population explosion, despite the civilizational achievements of the Neolithic revolution: sedentism
and agriculture. Over the subsequent five thousand years, by
contrast, world population would grow twentyfold, to more
than 100 million. The five thousand–year Neolithic transition was thus something of a demographic bottleneck, reflecting a nearly static level of reproduction. Supposing even
a population growth rate just barely over replacement levels
(for example, 0.015 percent) the total population would have

still more than doubled over these five millennia. One likely explanation for this paradox of apparent human progress in subsistence techniques together with long period of demographic stagnation is that, epidemiologically, this was perhaps the most lethal period in human history. In the case of Mesopotamia, the claim is that, owing precisely to the effects of the Neolithic revolution, it had become the focal point of chronic and acute infectious diseases that devastated the population again and again.[2]

Evidence in the archaeological record is hard to come by inasmuch as such diseases, unlike malnutrition, only rarely leave signature traces on human bones. Epidemic disease is, I believe, the "loudest" silence in the Neolithic archaeological record. Archaeology can assess only what it can recover and, in this case, we must speculate beyond the hard evidence. There are nonetheless good reasons for supposing that a great many of the sudden collapses of the earliest centers of population were due to devastating epidemic diseases.[3] Time and again there is evidence of a sudden and otherwise unexplained abandonment of previously well-populated sites. In the case of adverse climate change or soil salinization one would also expect depopulation, but in keeping with its cause it would be more likely to be regionwide and rather more gradual. Other explanations for the sudden evacuation or disappearance of a populous site are of course possible: civil war, conquest, floods. Epidemic disease, however, given the entirely novel crowding the Neolithic revolution made possible, is the most likely suspect, judging from the massive effects of disease that appear in the written records once they become available. The meaning of epidemic disease in this context is not confined to Homo sapiens alone. Epidemics affected domestic animals and crops that were also concentrated in the late-Neolithic multispecies resettlement camp. A population could as easily be devastated by a disease that swept through their flocks or their grain fields as by a plague that menaced them directly.

Once written records become available, however, we have ample evidence of deadly epidemics, which can, with caution, be read back to earlier periods. The Epic of Gilgamesh provides perhaps the most powerful evidence when its hero claims that his fame will outlive death as he depicts a scene of bodies felled, probably by pestilence, floating down the Euphrates. Mesopotamians, it seems, lived in the ever-threatening shadow of fatal epidemics. They had amulets, special prayers, prophylactic dolls, and "healing" goddesses and temples—the most famous of which was at Nippur—designed to ward off mass illness. Such events were, of course, poorly understood at the time. They were seen as "the devouring" of a god and as punishment for some transgression requiring compensatory ritual including the sacrifice of scapegoats.[4]

The first written sources also make it clear that early Mesopotamian populations understood the principle of "contagion" that spread epidemic disease. Where possible, they took steps to quarantine the first discernible cases, confining them to their quarters, letting no one out and no one in. They understood that long-distance travelers, traders, and soldiers were likely carriers of disease. Their practices of isolation and avoidance prefigured the quarantine procedures of the lazaretti of the Renaissance ports. An understanding of contagion was implicit not only in the avoidance of people who were infected but avoidance as well of their cups, dishes, clothes, and bed linen.[5] Soldiers returning from a campaign and suspected of carrying disease were obliged to burn their clothing and shields before entering the city. When isolation and quarantine failed, those who could fled the city, leaving the dying and deceased behind, and returning, if ever, only well after the epidemic had passed. In doing so, they must frequently have brought the epidemic to outlying areas, touching off a new round of quarantines and flight. There is little doubt in my mind that a good many of the earlier and unchronicled abandonments of populous areas were due more to disease than to politics.

Evidence for the role of pathogens in the diseases of humans, domesticated animals, and domesticate crops before the middle of the fourth millennium BCE is necessarily speculative. As written records proliferate, however, the evidence for epidemics grows in proportion; the texts refer, Karen Rhea Nemet-Nejat claims, to tuberculosis, typhus, bubonic plague, and smallpox.[6] One of the earliest and most amply attested is a devastating epidemic at Mari on the Euphrates in 1,800 BCE. The list of others is long, although the nature of the disease is typically obscure. The epidemic that destroyed the army of Sennachrib, son of Sargon II and Assyrian king in 701 BCE, that figures as well in the Old Testament's litany of plagues is now ascribed to typhus or cholera, the traditional scourges of armies on campaign. Later, the crushing plague of Athens in 430 BCE, described memorably by Thucydides, and the Antonine and Justinian plagues of Rome play a decisive role in what amounts to early "imperial" history. Given the larger populations and growing long-distance trade of this later era, there is little doubt that epidemics touched more

people and more areas than before. Nevertheless, Mesopotamia of the late fourth millennium BCE was a historically
novel environment for epidemics. By 3,200 BCE, Uruk was
the biggest city in the world, with anywhere from twenty-five
thousand to fifty thousand inhabitants, together with their
livestock and crops, dwarfing the concentrations of the earlier
Ubaid period. As the most demographically packed area, the
southern alluvium was especially vulnerable to epidemics; the
Akkadian word for epidemic disease "literally meant 'certain
death' and could be applied equally to animal as well as human
epidemics."[7] That concentration and an unprecedented flow
of trade created, as we shall now explain, a uniquely new vulnerability to the diseases of crowding.

Sedentism alone, well before widespread cultivation of
domesticated crops, created conditions of crowding that were
ideal "feedlots" for pathogens. The growth of large villages
and small towns in the Mesopotamian alluvium represented
a ten- to twentyfold increase in the population density over
anything Homo sapiens had previously experienced. The
logic of crowding and disease transmission is straightforward.
Imagine, for example, an enclosure with ten chickens, one of
which is infected with a parasite spread by droppings. After
a while—depending in part on the size of the enclosure, the
activity of the fowl, and the ease of transmission—another
chicken will become infected. Now, instead of ten chickens,
imagine five hundred chickens in the same enclosure and the
chances rise at least fiftyfold that another bird will become
quickly infected, and so on exponentially. Two birds are now
excreting the parasite, doubling the probability of a new infection. Recall that we have increased not only the poultry but

also their droppings by fifty times so that soon, the smaller the enclosure, the likelihood of other birds avoiding contact with the pathogen becomes vanishingly small.

For the present purposes we are applying the logic of crowding and diseases to Homo sapiens, but, as in the example above, it applies equally to the crowding of any disease-prone organism, flora or fauna. It is a crowding phenomenon that applies equally to flocks of birds and sheep, schools of fish, herds of reindeer or gazelle, and fields of cereals. The greater the genetic similarity—the less variation—the greater the likelihood that they will all be vulnerable to the same pathogen. Before extensive human travel, migratory birds that nested together combined long-distance travel with crowding to constitute, perhaps, the main vector for the spread of disease over distance. The association of infection with crowding was known and utilized long before the actual vectors of disease transmission were understood. Hunters and gatherers knew enough to stay clear of large settlements, and dispersal was long seen as a way to avoid contracting an epidemic disease. Late medieval Oxford and Cambridge maintained plague houses in the countryside to which students were dispatched with the first sign of the plague. Concentration could be lethal. Thus the trenches, demobilization camps, and troop ships at the conclusion of World War I provided the ideal conditions for the massive and lethal influenza pandemic of 1918. Social sites of crowding—fairs, military encampments, schools, prisons, slums, religious pilgrimages, such as the hajj to Mecca—have historically been locations where infectious diseases have been contracted and from which they have subsequently been dispersed.

The importance of sedentism and the crowding it allowed can hardly be overestimated. It means that virtually all the infectious diseases due to microorganisms specifically adapted to Homo sapiens came into existence only in the past ten thousand years, many of them perhaps only in the past five thousand. They were, in the strong sense, a "civilizational effect." These historically novel diseases — cholera, smallpox, mumps, measles, influenza, chicken pox, and perhaps malaria — arose only as a result of the beginnings of urbanism and, as we shall see, agriculture. Until very recently they collectively represented the major overall cause of human mortality. It is not as if presedentary populations did not have their own parasites and diseases, but such diseases would have been not the crowding diseases but rather diseases characterized by long latency and/or a nonhuman reservoir: typhoid, amoebic dysentery, herpes, trachoma, leprosy, schistosomiasis, filariasis.[8]

The diseases of crowding are also called density-dependent diseases or, in contemporary public health parlance, acute community infections. For many viral diseases that have come to depend on a human host, it is possible, by knowing the mode of transmission, the duration of infectivity, and the duration of acquired immunity after infection, to infer the minimal population required to keep the infection from dying out for lack of new hosts. Epidemiologists are fond of citing the example of measles in the isolated Faroe Islands in the eighteenth and nineteenth centuries. An epidemic brought by sailors devastated the islands in 1781, and, given the lifelong immunity conferred on survivors, the islands were free of the measles for sixty-five years until 1846, when it returned, infecting all but the aged folks who had survived the earlier epidemic. A further epidemic thirty years later infected only those under thirty. For measles specifically, epidemiologists have calculated that at least 3,000 newly susceptible hosts would be required annually to sustain a permanent infection and that only a population of roughly 300,000 could provide this many hosts. Having a population far below this threshold, the Faroe Islands had to "import" its measles anew for each epidemic. By the same token, of course, this means that none of these diseases could have existed before the populations of the Neolithic. It also explains the generally vibrant good health of the New World populations—as well as their later vulnerability to the Old World pathogens. The groups crossing the Bering Strait in several waves around 13,000 BCE came before most such diseases had arisen and, in any case, in groups far too small to sustain any of the crowding diseases.

No account of the epidemiology of the Neolithic is complete without noting the key role of domesticates: livestock, commensals, and cultivated grains and legumes. The key principle of crowding is again operative. The Neolithic was not only an unprecedented gathering of people but, at the same time, a wholly unprecedented gathering of sheep, goats, cattle, pigs, dogs, cats, chickens, ducks, geese. To the degree that they were already "herd" or "flock" animals, they would have carried some species-specific pathogens of crowding. Assembled for the first time around the domus, in close and continuous contact, they quickly came to share a wide range of infective organisms. Estimates vary, but of the fourteen hundred known human pathogenic organisms, between eight hundred and nine hundred are *zoonotic* diseases, originating in nonhuman hosts. For most of these pathogens, Homo sapiens is a final "dead-end" host: humans do not transmit it further to another nonhuman host.

The multispecies resettlement camp was, then, not only a historic assemblage of mammals in numbers and proximity never previously known, but it was also an assembly of all the bacteria, protozoa, helminthes, and viruses that fed on them. The victors, as it were, in this pest race were those pathogens that could quickly adapt to new hosts in the domus and multiply. What was occurring was the first massive surge of pathogens across the species barrier, establishing an entirely new epidemiological order. The narrative of this breach is naturally told from the (horrified) perspective of Homo sapiens. It cannot have been any less melancholy from the perspective of, say, the goat or sheep that, after all, did not volunteer to enter the domus. I leave it to the reader to imagine how a precocious, all-knowing goat might narrate the history of disease transmission in the Neolithic.

The list of diseases shared with domesticates and commensals at the domus is quantitatively striking. In an outdated list, now surely even longer, we humans share twenty-six diseases with poultry, thirty-two with rats and mice, thirty-five with horses, forty-two with pigs, forty-six with sheep and goats, fifty with cattle, and sixty-five with our much-studied and oldest domesticate, the dog.[9] Measles is suspected to have arisen from a rinderpest virus among sheep and goats, smallpox from camel domestication and a cowpox-bearing rodent ancestor, and influenza from the domestication of waterfowl some forty-five hundred years ago. The generation of new species-jumping zoonoses grew as populations of man and beasts swelled and contact over longer distances became more frequent. It continues today. Little wonder, then, that southeast China, specifically Guangdong, probably the largest, most crowded, and historically deepest concentration of Homo sapiens, pigs, chickens, geese, ducks, and wild animal markets in the world, has been a major world petri dish for the incubation of new strains of bird and swine flu.

The disease ecology of the late Neolithic was not simply a result of the crowding of people and their domesticates in fixed settlements. It was rather an effect of the entire domus complex as an ecological module. The clearing of the land for agriculture and the grazing of the new domesticates created an entirely new landscape, and an entirely new ecological niche with more sunlight, more exposed soils, into which new suites of flora, fauna, insects, and microorganisms moved as the previous ecological pattern was disturbed. Some of the transformation was by design, as with crops, but much more represented the second- and third-order collateral effects of the domus's invention.

Emblematic of this collateral effect was the concentration of animal and human wastes: in particular, feces. The relative immobility of sedentary humans and livestock and their wastes permits repeated infection with the same varieties of parasites. Mosquitoes and arthropods, often the vectors of disease, find the wastes ideal sites for breeding and feeding. Mobile groups of hunter-gatherers, by contrast, often leave their parasites behind by moving to a new environment where they cannot breed. Once stationary, the domus, with its humans, livestock, grain, feces, and plant wastes, makes an attractive feedlot for many commensals, from rats and swallows down the chain of predation to fleas and lice,

bacteria and protozoa. The pioneers who created this historically novel ecology could not possibly have known the disease
vectors they were inadvertently unleashing. In fact, it was not
until the late nineteenth-century discoveries of the founders
of microbiology, Robert Koch and Louis Pasteur, that it became clear what a heavy price in chronic and lethal infections
Homo sapiens was paying for the absence of clean water, sanitation, and sewage removal. As devastating new illnesses left
humans not knowing what hit them, folk theories and reme-
dies proliferated. Only one nostrum — "dispersal" — implicitly
identified crowding as the basic cause.

The density-dependent diseases afflicting the populations
of the late-Neolithic multispecies resettlement camp represented a new and rigorous selection pressure from pathogens
never experienced by their ancestors. One imagines that not
a few early concentrations of sedentary peoples were all but
exterminated by diseases to which they had virtually no resistance. For smaller preliterate societies it is all but impossible to know for sure the role of epidemics in mortality, and
much of the evidence from early cemeteries in inconclusive. It
is quite likely, however, that the crowding diseases, including
especially zoonoses, were largely responsible for the demographic bottleneck of the early Neolithic. In time — how long
is uncertain and varies with the pathogen — crowded populations developed a degree of immunity to many pathogens,
which in turn became endemic, signifying a stable and less
lethal pathogen-host relationship. After all, only those who
survive live on to have children! Some diseases — whooping
cough and meningitis, for example — might still endanger the
very young, while others, if contracted by a younger young

person, were relatively harmless and conferred immunity: polio, smallpox, measles, mumps, and infectious hepatitis.[10]

Once a disease becomes endemic in a sedentary population, it is far less lethal, often circulating largely in a subclinical form for most carriers. At this point, unexposed populations having little or no immunity against this pathogen are likely to be uniquely vulnerable when they come into contact with a population in which it is endemic. Thus war captives, slaves, and migrants from distant or isolated villages previously outside the circle of crowd immunity have fewer defenses and are likely to succumb to diseases to which large sedentary populations have become, over time, largely immune. It was for this reason, of course, that the encounter between the Old World and the New World proved so cataclysmic for the immunologically naïve Native Americans, isolated for more than ten millennia from Old World pathogens.

The diseases of sedentism and crowding in the late Neolithic were compounded by an increasingly agricultural diet, deficient in many essential nutrients. One's chances of surviving an epidemic disease, other things equal, especially as an infant or a pregnant woman, depended very much on one's nutritional status. The extremely high rates of mortality for infants (40–50 percent) among most early agriculturalists was a result of the conjuncture of a diet that weakened the vulnerable with new infectious diseases that carried them off.

Evidence for the relative restriction and impoverishment of early farmers' diets comes largely from comparisons of skeletal remains of farmers with those of hunter-gatherers living nearby at the same time. The hunter-gatherers were several inches taller on average. This presumably reflected their more varied and abundant diet. It would be hard, as we have explained, to exaggerate that variety. Not only might it span several food webs—marine, wetland, forest, savanna, arid— each with its seasonal variation, but even when it came to plant foods, the diversity was, by agricultural standards, staggering. The archaeological site of Abu Hureyra, for example, in its hunter-gatherer phase, yielded remains from 192 different plants, of which 142 could be identified, and of which 118 are known to be consumed by contemporary hunter-gatherers.[11]

A symposium devoted to assessing the impact of the Neolithic revolution on human health worldwide concluded on the basis of paleopathological data:

> [Nutritional] stress . . . does not seem to have become common and widespread until after the development of high degrees of sedentism, population density, and reliance on agriculture. At this stage . . . the incidence of physiological stress increases greatly and the average mortality rates increase appreciably. Most of these agricultural populations have high frequencies of *porotic hyperostosis* [overgrowth of poorly formed bone associated with malnutrition, particularly iron-deficiency related malnutrition] and *cribra orbitalia* [a localized version of the above condition, in the eye socket], and there is a substantial increase in the number and severity of [tooth] enamel *hypoplasis* and pathologies associated with infectious diseases.[12]

Much of the malnutrition detected in what we might call "agricultural woman"—for women, owing to blood loss with menses, were the most severely affected—seems to be due to iron deficiency. Preagricultural women had a diet that supplied abundant amounts of omega-6 and omega-3 fatty acids derived from game, fish, and certain plant oils. These fatty acids are important because they facilitate the uptake of iron necessary for the formation of oxygen-carrying red blood cells. Cereal diets, by contrast, not only lack the essential fatty acids but actually inhibit the uptake of iron. The result of the first increasingly intensive cereal diets in the late Neolithic (wheat, barley, millet) was therefore the appearance of iron-deficiency anemia, leaving an unmistakable forensic bone signature.

Most of the added vulnerability to novel infections seems due to a relatively high and narrow carbohydrate diet without much in the way of wild foods and meat. It was likely to lack some essential vitamins and to be protein poor. Even the meat of the domesticates on which they might occasionally feast contained far fewer vital fatty acids than wild game. Illnesses attributable to the Neolithic diet that do have bone signatures, such as rickets, can be documented; those that affect the soft tissues are far harder to document (except in the occasional well-preserved mummy). Nevertheless, on the basis of dietary knowledge and early written accounts of illnesses that can probably be assumed, again on dietary knowledge, to have existed earlier, the following nutrition-related diseases have been attributed to Neolithic foodways: beriberi, pellagra, riboflavin deficiency, and kwashiorkor.

What about crops? They too were subjected to a kind of "sedentism" on fixed fields and conditions of crowding, as well as a new, human-driven selection process that reduced their genetic diversity to foster desired characteristics. They too, like any organism, were subject to their own density-dependent diseases, as we shall see. Because "both herding and agriculture are frequently afflicted with epidemics, crop failure, or other misfortunes," Nissen and Heine claim that early farmers preferred, when possible, to rely on hunting, fishing, and gathering.[13] Here again the archaeological record is not very helpful. It is possible to show, say, that a previously populous area was suddenly abandoned; before written records, however, knowing why it was deserted is another matter. A crop fungus, a rust, an insect infestation, or even a storm that destroys a ripe crop, like soft-tissue diseases, leave little or no trace. Written records, when they are available, are more likely to record a "harvest failure" or famine than to specify the cause, which, in many cases, is not understood by the victims themselves.

Crops represented their own perfect "floral" epidemiological storm. Consider as a pathogen or insect might the attractions of the Neolithic agricultural landscape. It was not only crowded but, compared with wild grasslands, was largely devoted to just two major grains: wheat and barley. Furthermore, these were fixed fields cropped more or less continuously, as compared, say, with fire-field cultivation (aka swidden or slash-and-burn), where a field was planted for a year or two and then fallowed for a decade or more. Repeated annual cultivation provided, in effect, a permanent feedlot for insect pests and plant diseases—not to mention obligate weeds—which built up to population levels that could not have existed before fixed-field monocropping. Large sedentary communities necessarily meant many arable fields in close proximity, growing a similar variety of crop; this promoted a commensurate buildup of pest populations. As with the epidemiology of human crowding, it seems logical to suppose that many of the crop diseases besetting Neolithic planters were new pathogens that evolved to take advantage of such a nutritious agro-ecology. The literal meaning of "parasite," from the original Greek root, is "beside the grain."

Crops not only are threatened, as are humans, with bacterial, fungal, and viral diseases, but they face a host of predators large and small—snails, slugs, insects, birds, rodents, and other mammals, as well as a large variety of evolving weeds that compete with the cultivar for nutrition, water, light, and space.[14] The seed in the ground is attacked by insect larvae, rodents, and birds. During growth and grain development the same pests are still active, as well as aphids that suck sap and transmit disease. Fungal diseases are especially devastating, including mildew, smut, bunt, rusts, and ergot (famous as St. Anthony's Fire when ingested by humans) at this stage. The part of the crop that does not succumb to these predators must compete with a host of weeds that have come to specialize in ploughed soil and to mimic certain crops. And once the harvest is in the granary it is still subject to weevils, rodents, and fungi.

It is common enough in the contemporary Middle East for several crops in succession to be lost to insects, birds, or disease. In an experiment in northern Europe, a crop of modern barley, fertilized but not protected with modern herbicides or pesticides, was reduced by half: 20 percent due to crop disease, 12 percent to animals, and 18 percent to weeds.[15] Threatened by the diseases of crowding and monoculture, domesticated crops must be constantly defended by their human custodians if they are to yield a harvest. It is largely for this reason that early agriculture was so dauntingly labor intensive. Various techniques were devised to reduce the labor involved and improve the yields. Fields were scattered so that they were less contiguous; fallowing and crop rotation was practiced; and seed was procured at a distance to reduce genetic uniformity. Ripening crops were closely guarded by farmers, their families, and scarecrows. But given the disease-prone agroecology of the domesticated crop, it was touch and go whether the crop would survive all the predators to feed its ultimate guardian and predator: the farmer.

The older narrative of civilizational progress is, in one basic respect, undoubtedly correct. The domestication of plants and animals made possible a degree of sedentism that did form the basis of the earliest civilizations and states and their cultural achievements. It rested, however, on an extremely slender and fragile genetic foundation: a handful of crops, a few species of livestock, and a radically simplified landscape that had to be constantly defended against a reconquest by excluded nature. At the same time, the domus was never even remotely self-sufficient. It required a constant subsidy, as it were, from that excluded nature: wood for fuel and building, fish, mollusks, woodland grazing, small game, wild vegetables, fruits, and nuts. In a famine, farmers resorted to all the extradomus resources that hunter-gatherers relied on.

The domus was at the same time a veritable feast and a pilgrimage site for uninvited commensals and pests large and small, down to the smallest viruses. Its very concentration and simplicity made it uniquely vulnerable to collapse. Late Neolithic agriculture was the first of many steps in the development of special techniques for maximizing the production of a small number of preferred plant and animal species. An illness—of crops, livestock, or people—a drought, excessive rains, a plague of locusts, rats, or birds, could bring the whole edifice down in the blink of an eye. Based on a narrow food web, Neolithic agriculture was far more productive, in a concentrated way, but also far more fragile than hunting and gathering or even shifting-cultivation, which combined mobility with a reliance on a diversity of foods. How, despite its fragility, the domus module of fixed-field agriculture became a hegemonic, agro-ecological and demographic bulldozer that transformed much of the world in its image is something of a miracle.

A NOTE ON FERTILITY AND POPULATION

The ultimate dominance of the Neolithic grain complex is hardly prefigured by the epidemiology of the domus. An attentive reader might not only be puzzled by the rise of agrarian civilization but might wonder how, in light of the pathogens Neolithic cultivators faced, this new form of agrarian life managed to survive at all, let alone thrive.

The short answer, I believe, is sedentism itself. Despite general ill health and high infant and maternal mortality vis-à-vis hunters and gatherers, it turns out that sedentary agriculturalists also had unprecedentedly high rates of reproduction—enough to more than compensate for the also unprecedentedly high rates of mortality. The effect of the transition to sedentism on fertility has been convincingly documented in contemporary studies by Richard Lee, comparing newly settled with still-mobile !Kung Bushman women, as well as other studies making more comprehensive comparisons of fertility between farmers and foragers.[16]

Nonsedentary populations typically limit their reproduction deliberately. The logistics of moving camp regularly make it burdensome, if not impossible, to have two infants who must be carried at the same time. As a result, the spacing of children of hunter-gatherers is on the order of four years, a spacing that is achieved by delayed weaning, abortifacients, and neglect or infanticide. Furthermore, some combination of strenuous exercise with a lean and protein-rich diet meant that puberty arrived later, ovulation was less regular, and menopause arrived earlier. Among sedentary agriculturalists, by contrast, the burden of a much shorter spacing of children as experienced by mobile foragers is much reduced and, as we shall see, the greater value of the children as a labor force in agriculture is enhanced. By virtue of sedentism, menarche is earlier; with a grain diet, infants can be weaned earlier on soft foods; and by virtue of a high-carbohydrate diet, ovulation is encouraged and a woman's reproductive life is extended.

Given the disease burden of agrarian society and its fragility, the demographic "advantage" of farmers over hunter-gatherers might have been quite small. But the thing to remember in this context is that over a period of five thousand years—like the "miracle" of compound interest—the eventual difference became massive. For example, if one computes doubling times for different rates of reproduction, it turns out that an annual rate of 0.014 percent doubles population in five thousand years while a rate of 0.028 percent, still minuscule, doubles population in half that time (twenty-five hundred years), and, of course, doubles again to a total four times as great after five thousand years. Given enough time, the small reproductive advantage of farmers was overwhelming.[17]

The demographic expansion (if the crude order of magnitude we are using is realistic) of world population from four million to five million over five thousand years seems puny indeed. As the proportion of Neolithic farmers to hunter-gatherers was far greater in 5,000 BCE than in 10,000 BCE, it is quite likely that even in this bottleneck period, the grain farmers of the world were demographically overtaking hunter-gatherers. The two other possibilities are that many hunter-gatherers were taking up agriculture by choice or force or that the agrarian pathogens that had become endemic and less lethal to farmers were devastating the still immunologically naïve hunter-gatherers with whom they came into contact, much as European pathogens killed a great majority of the New World's population.[18] There is no clear evidence to confirm or reject these possibilities. One way or another, however, Neolithic farming communities in the Levant, Egypt, and China were expanding and spreading to alluvial bottomlands, apparently at the expense of nonsedentary peoples. The writing, however faint, was on the wall.

CHAPTER FOUR

Agro-ecology of the Early State

Whoever has silver, whoever has jewels, whoever has cattle, whoever has sheep shall take a seat at the gate of whoever has grain, and pass his time there.
—Sumerian text: Debate between Sheep and Grain

Ultimately men bow down to the man, or group of men, who can and dare take over the hoard, the store of bread, the riches, to distribute among the people again.
—D. H. Lawrence

IF civilization is judged an achievement of the state, and if archaic civilization means sedentism, farming, the domus, irrigation, and towns, then there is something radically wrong with the historical order. All of these human achievements of the Neolithic were in place well before we encounter anything like a state in Mesopotamia. Quite the contrary. On the basis of what we now know, the embryonic state arises by harnessing the late Neolithic grain and manpower module as a basis of control and appropriation. The module was, as we shall see, the only possible scaffolding available for the design of a state.

Settled populations growing crops of domesticated grains, and small towns with a thousand or more inhabitants facilitating commerce, were an autonomous achievement of the Neolithic, being in place nearly two millennia before the appearance of the first states, around 3,300 BCE.[1] These earliest towns are, Jennifer Pournelle reminds us, "better imagined as islands embedded in a marshy plain, situated on the borders and in the heart of vast deltaic marshlands." "Their waterways served less as irrigation canals than as transportation routes."[2] Although there were earlier proto-urban settlements elsewhere in the region outside the southern alluvium, it seems clear that urbanism, thanks to wetland abundance, was more persistent, durable, and resilient in the alluvium than anywhere else.[3]

This complex, however, represented a unique new concentration of manpower, arable land, and nutrition that, if "captured"—"parasitized" might not be too strong a word—could be made into a powerful node of political power and privilege. The Neolithic agro-complex was a necessary but not a sufficient basis for state formation; it made state formation possible but not certain. In Weberian terms, we are dealing here with something like "elective affinity" rather than cause and effect. Thus it was possible and not uncommon at the time to have sedentary farming populations on alluvial soils practicing irrigation without any state.[4] But there was no such thing as a state that did not rest on an alluvial, grain-farming population.

What constitutes a state in this context? How would we know the first pristine state when we saw it? The answer is not cut and dried; I am inclined to see "stateness" as a more-or-less proposition rather than strictly either/or. There are many plausible attributes to stateness, and the more of them a particular polity possesses, the more likely we are to call it a state. Small embryonic towns of sedentary foragers, cultivators, and pastoralists that manage their collective affairs and trade with the outside world are not, ipso facto, states. Nor is the standard Weberian criterion of a territorial political unit that monopolizes the application of coercive force entirely adequate, for it takes so many other features of states for granted. We think of states as institutions that have strata of officials specialized in the assessment and collection of taxes—whether in grain, labor, or specie—and who are responsible to a ruler or rulers. We think of states as exercising executive power in a fairly complex, stratified, hierarchical society with an appreciable division of labor (weavers, artisans, priests, metalworkers, clerks, soldiers, cultivators). Some would apply more stringent criteria: a state should have an army, defensive walls, a monumental ritual center or palace, and perhaps a king or queen.[5]

Pinpointing the birth of the early state, given these various attributes, is a relatively arbitrary exercise that is further constrained by the few sites from which we have convincing archaeological and historical evidence. Among these characteristics, I propose to privilege those that point to territoriality and a specialized state apparatus: walls, tax collection, and officials. By such standards there is no doubt that the "state" of Uruk is firmly in place by 3,200 BCE. Nissen calls the period from 3,200 to 2,800 BCE the "era of high civilization" in the Near East, during which "Babylonia was, without doubt, the region that produced the most complex economic, political and social orders."[6] Not incidentally, the iconic founding act of establishing a Sumerian polity was the building of a city wall. A wall at Uruk was, in fact, built between 3,300 and 3,000 BCE, when Gilgamesh was thought by some to have reigned. Uruk was the pioneer of the state form that would be replicated throughout the Mesopotamian alluvium by roughly twenty other competing city-states or "peer polities." These polities were small enough that one could walk from the center of most to the outer boundary in a day.

With political and economic dominance over a modest agricultural hinterland, as well as a structured city government, the Sumerian city of Uruk in the late fourth millennium BCE met the criteria of the city-state. It was, at first, unique in its size and power. We have enough evidence to demonstrate, however, that by the first half of the third millennium, at the latest, major cities such as Kish, Nippur, Isin, Lagash, Eridu, and Ur belong to the same category as Uruk.[7]

If Uruk looms particularly large in this and other examinations of early state making, it is not simply because it seems to be the first state but because it is, at the same time, the most documented archaeologically. Compared with Uruk, our knowledge of other early state centers in Mesopotamia is fragmentary. For its time, it was almost surely the largest city in the world in both physical extent and in population. Estimates of its population range from twenty-five thousand to fifty thousand; the number of inhabitants tripled over two hundred years, an increase unlikely to have come from natural population growth, given the high mortality rates. As the place-names of Ur, Uruk, and Eridu appear not to be of Sumerian origin, this suggests an in-migration displacing or absorbing earlier inhabitants. The bas reliefs depicting prisoners of war in neck shackles suggest another means by which the population was augmented.

Uruk's walls appear to have enclosed an area of 250 hectares, twice the size of classical Athens nearly three millennia later. Given Postgate's calculation that another Sumerian city, Abu Salabikh, with its hypothetical population of about ten thousand, would have had to dominate a rural hinterland for ten kilometers around, one imagines that Uruk's hinterland would have been at least two or three times as great.[8] There is, moreover, abundant evidence of substantial work gangs mobilized for agricultural and nonagricultural tasks by temples, as well as thousands of standardized bowls used, most judge, to distribute food or beer rations. Other marks of stateness include a specialist scribal class, soldiers (full-time?) with armor, and efforts at standardizing weights and measures. Most of my discussion of the early state, therefore, unless otherwise noted, relies on the extensive literature on Uruk with occasional references to the nearby, well-documented but short-lived Third Dynasty of Ur (Ur III) a millennium later.

If state formation depends on the control, maintenance, and expansion of the concentrations of grain and manpower on the alluvium, the question arises of how the early state could have come to dominate these population-and-grain modules. The would-be subjects of this hypothetical state, after all, had direct, unmediated access to water and flood-retreat agriculture as well as a variety of subsistence options beyond cultivation. One convincing explanation for how this cultivating population might have been assembled as state subjects is climate change. Nissen shows that the period from at least 3,500 to 2,500 BCE was marked by a steep decline in sea level and a decline in the water volume in the Euphrates. Increasing aridity meant that the rivers shrank back to their main channels and the population increasingly huddled around the remaining watercourses, while soil salinization of water-deprived areas sharply reduced the amount of arable land. In the process, the population became strikingly more concentrated, more "urban." Irrigation became both more important and more labor intensive—it now often required lifting water—and access to dug canals became vital. City states (for example, Umma and Lagash) fought over arable land and access to the water that could irrigate it. Over time a more reticulated canal system dug with corvée or slave labor developed. If Nissen's scenario of aridity and its demographic consequence of concentration, both of which rest on solid evidence, is accepted, it provides one plausible account of state formation. The shortage of irrigation water confined the population increasingly to well-watered places and eliminated or diminished many of the alternative form of subsistence, such as foraging and hunting. As Nissen describes it, "We have already seen this happening in the previous period, where the tendency began to emerge for settlements to concentrate around the courses of the larger rivers, while the area between the rivers became increasingly empty."[9] Climate change, then, by forcing a kind of urbanization in which 90 percent of the population lived in settlements of thirty hectares or more, intensified the grain-and-manpower modules that were ideal for state formation. Aridity proved the indispensable handmaiden of state making by delivering, as it were, an assembled population and concentrated cereal grains in an embryonic state space that could not, at that epoch, have been assembled by any other means.

Not just in Mesopotamia but virtually everywhere, it seems, early state battens itself onto this new source of sustenance. The dense concentration of grain and manpower on the only soils capable of sustaining them in such numbers—alluvial or loess soils—maximized the possibilities of appropriation, stratification, and inequality. The state form colonizes this nucleus as its productive base, scales it up, intensifies it, and occasionally adds infrastructure—such as canals for transport and irrigation—in the interest of fattening and protecting the goose that lays the golden eggs. In terms used earlier, one can think of these forms of intensification as elite niche-construction: modifying the landscape and ecology so as to enrich the productivity of its habitat. It is, of course, only in the context of rich soils and available water that the ecological capacity for the further intensification of agriculture and population growth was possible, and thus it was only in such settings that the first bureaucratic states were likely to arise.

The development of the Mesopotamian state was not remotely linear. Statelets in the alluvium had, like their inhabitants, a very short life expectancy. Interregna were more common than "regna," and long episodes of collapse and disintegration were commonplace. As we have seen, the late Neolithic proto-urban complex was a touch-and-go affair under the best of circumstances. It was menaced by variable rainfall, floods, pest attacks, and any number of crop, livestock, and human diseases that could wipe out a settlement or, more likely, force its residents to scatter as hunters, foragers, and pastoralists so as to sustain themselves.

To the already considerable perils of the crowded Neolithic complex, the superimposition of the state added an additional layer of fragility and insecurity. Taxes and warfare can serve to illustrate the added fragility. Taxes in kind (grain or livestock) or in labor obviously meant that the farmer was not only producing for the domus but had to supply a fund of rent that elites appropriated for their own subsistence and display, although the same elites might occasionally disburse stored grain in a famine to keep their population intact. It is hard to determine how burdensome this tax was, and in any case, it varied over time and between polities. To judge from agrarian history in general, the tax in grain is unlikely to have been less than a fifth of the harvest. Cultivators walked, in effect, closer to the subsistence precipice: a crop failure that, without taxes, might mean hunger could, after the state took its taxes, mean utter ruin.

The evidence for frequent warfare among rival polities in the southern alluvium is abundant. It is hard to tell precisely how sanguinary it was, but given the preciousness of population for all the early states, wars were probably more destructive than bloody. One account of warfare among the peer polities of the alluvium asserts that the population lived at the subsistence level except when a victorious army returned with loot and tribute.[10] The gains of the winner were offset by the losses of the vanquished. Warfare itself meant the burning of crops, the seizure of granaries, the confiscation of livestock and household goods—one's own army was as likely to be as big a threat to livelihood as the enemy's. The early state, rather like the weather, was more often an added threat to subsistence than its benefactor.

THE AGRO-GEOGRAPHY OF STATE-MAKING

Archaic states, in the crudest material terms, were all agrarian and required an appropriable surplus of agro-pastoral products to feed nonproducers: clerks, artisans, soldiers, priests, aristocrats. Given the logistics of transport in the ancient world, this meant the concentration of as much arable land and as many people to work it as possible within the smallest radius. The late-Neolithic resettlement camp located on rich alluvial soil was the already existing nucleus of people and grain from which a state could be elaborated.

We can be more specific about the geographical conditions for state building. Only the richest soils were productive enough per hectare to sustain a large population in a compact area and to produce a taxable surplus. In practice this meant loess (wind deposited) or alluvial (flood deposited) soils. Alluvia, the historic gift of the annual floods of the Tigris and Euphrates and their tributaries, were the sites of state making in Mesopotamia: no alluvium, no state.[11] If reliable and noncatastrophic floods allowed, flood-retreat agriculture could be practiced on the easily worked and nutritious silt (in Egypt along the Nile as well), in which case the density of the population might be even greater. Much the same can be said for the earliest state centers in China (Qin and Han Dynasties), in the loess soils along the Yellow River, where population density reached levels rare for preindustrial societies. To follow the progress of the Chinese state is to follow the agro-ecology that made it possible. As Owen Lattimore noted, "Irrigation was spectacularly rewarding in the loess core of ancient China, soft, easily worked soil, no stone, a climate allowing many different crops—the complex spread farther and farther out so long as land was suitable."[12]

Water, of course, was vital. Its abundance in the wetlands provided, as we have seen, the basis for some of the first substantial sedentary communities. Only well-watered alluvium, whether by reliable rainfall or irrigation water close at hand, was a possible site for state making. But water was vital in other ways as well. Located at or near a floodplain and specializing in grain agriculture, none of the early state centers in Mesopotamia was even remotely self-sufficient economically. They required a host of products that originated in other ecological zones: timber, firewood, leather, obsidian, copper, tin, gold and silver, and honey. In exchange, the small statelets might trade pottery, cloth, grain, and artisanal products.[13] Most of these goods had to move by water rather than overland. I am tempted to say, "no water transport, no state"—only a slight exaggeration.[14] We have already emphasized earlier how transportation by ship or small barge is exponentially more economical than shipment by donkey or cart. Illustrating the contrast is the striking fact that as late as 1800 (before the steamship or railroad) it was about as fast to go from Southampton, England, to the Cape of Good Hope by ship as it was to go by stagecoach from London to Edinburgh.[15] And of course, the ship could carry vastly more cargo. The miracle of eliminating so much friction by water transport has meant that it was a very rare early state that did not depend on nearby navigable waterways—coastal or riverine— to trade for its requirements. Being located near the bottom of the Tigris-Euphrates watershed, the earliest alluvial states could also take advantage of the current to float bulk com- modities such as timber, with minimum expenditure of labor. It is perhaps no coincidence that in the middle passages of the Epic of Gilgamesh is a narrative of floating a raft of cedar— which will become the main gate of the newly founded city— down the river after killing the giant guarding the great forest.

Avoiding friction in general is important to state making. Navigable, calm water for much of the year is typically essen- tial. It helps if the land is flat, as well. A floodplain is basically flat by definition, while rugged terrain adds, again exponen- tially, to the cost of transport. Grasping the implicit ecology of state formation, Ibn Khaldun noted that the Arabs could conquer lands that were flat but were stymied by mountains and ravines.[16]

Specifying the conditions of elementary state making helps us appreciate the obverse: the conditions under which state formation is unlikely or indeed impossible. As the con- centration of population facilitates state making, dispersal thwarts it. Because it is the rich, well-watered alluvium that allows for such concentration, it follows that nonalluvium ecologies are unlikely to be sites of early states. Arid deserts and mountainous zones (barring fertile intermontane basins) virtually require dispersed subsistence strategies and can hardly serve as the nucleus of a state. These "nonstate spaces," owing to their different subsistence patterns and social orga- nization—pastoralism, foraging, and slash-and-burn cultivation—are often stigmatized and coded "barbarian" by state discourses.

The state "module" requires concentrated manpower—specifically agricultural manpower practicing mainly fixed-field cultivation. Concentration alone will not do. The wetlands ecology of the southern part of the Mesopotamian alluvium, where substantial sedentism first arose in the Middle East, is a case in point.[17] It was heavily populated and although some crops were grown, its earliest towns yield no remains of the regular ploughed fields that leave an unmistakable signature in the archaeological record. Livelihoods here, as described earlier, were exceptionally diverse: wetland foraging and hunting, harvesting wild reeds and sedges, recessional grazing of sheep, goats, and cattle. Despite a dense and affluent population, this was not an agricultural population. "Rather than supporting a model of social transformation driven by irrigated grain crops, this revisualized heartland of cities suggests a settlement progression beginning with . . . opportunistic dependence on littoral bio-mass."[18] The wetlands produced wealth and towns but no states until more than a millennium later. Unlike a landscape of plough agriculture, the exuberant diversity of livelihoods in the wetlands was not favorable to state making. As if to confirm the suspicion that larger river deltas are not conducive to early state building, the Nile Delta seems to provide a comparable case. Early Egyptian states arose upriver from the Delta, which, though also well populated and rich in subsistence resources, was not the basis of a state. On the contrary, it was seen as a zone of hostility and resistance to the state. Like the inhabitants of the Mesopotamian wetlands, the Nile Delta population lived on turtlebacks, fished, harvested reeds, ate shellfish, and practiced little if any agriculture; they were not a part of dynastic Egypt.

The heartland of early states along the Yellow River were, similarly, upriver and not in the turbulent, ever-changing delta area. Cultivation, though it was of millet, was as vital to the state-building nucleus in China as wheat and barley were to the Mesopotamian states. The Chinese state-building project, hopped, as it were, from one rich arable loess location to another, leaving aside both the hilly blocks of land ("inner" barbarians) between them and the complex, diverse Yellow River Delta.

GRAINS MAKE STATES

The subsistence bases of all the earliest, major agrarian states of antiquity—Mesopotamia, Egypt, Indus Valley, Yellow River—bear a remarkable resemblance to one another. They are all grain states: wheat, barley, and, in the case of the Yellow River, millet. Subsequent early states follow the same pattern, although irrigated rice and, in the New World, maize are added to the list of staple crops. A partial exception to this rule might be the Inka state, which relied on maize and potatoes, although maize seems to have predominated as the tax crop.[19] In a grain state, one or two cereal grains provided the main food starch, the unit of taxation in kind, and the basis for a hegemonic agrarian calendar. Such states were confined to the ecological niches where alluvial soils and available water made them possible. Here the emphasis should be again on

Lucien Febvre's concept of "possibilism"; such a niche was necessary for state formation (and could be expanded by landscape management such as canals and terracing), but it was not sufficient.[20] And in this case, population concentration must be distinguished from state making; wetlands abundance, as we have seen, could lead to incipient urbanism and commerce, but did not lead to state formation without grain growing on a large scale.[21]

Why, however, should cereal grains play such a massive role in the earliest states? After all, other crops, in particular legumes such as lentils, chickpeas, and peas, had been domesticated in the Middle East and, in China, taro and soybean. Why were they not the basis of state formation? More broadly, why have no "lentil states," chickpea states, taro states, sago states, breadfruit states, yam states, cassava states, potato states, peanut states, or banana states appeared in the historical record? Many of these cultivars provide more calories per unit of land than wheat and barley, some require less labor, and singly or in combination they would provide comparable basic nutrition. Many of them meet, in other words, the agro-demographic conditions of population density and food value as well as cereal grains. Only irrigated rice outclasses them in terms of sheer concentration of caloric value per unit of land.[22]

The key to the nexus between grains and states lies, I believe, in the fact that only the cereal grains can serve as a basis for taxation: visible, divisible, assessable, storable, transportable, and "rationable." Other crops—legumes, tubers, and starch plants—have some of these desirable state-adapted qualities, but none has all of these advantages. To appreciate the unique advantages of the cereal grains, it helps to place yourself in the sandals of an ancient tax-collection official interested, above all, in the ease and efficiency of appropriation.

The fact that cereal grains grow above ground and ripen at roughly the same time makes the job of any would-be taxman that much easier. If the army or the tax officials arrive at the right time, they can cut, thresh, and confiscate the entire harvest in one operation. For a hostile army, cereal grains make a scorched-earth policy that much simpler; they can burn the harvest-ready grain fields and reduce the cultivators to flight or starvation. Better yet, a tax collector or enemy can simply wait until the crop has been threshed and stored and confiscate the entire contents of the granary. In practice, in the case of the medieval tithe, the cultivator was expected to assemble the unthreshed grain in sheaves in the field, from which the tithe collector would take every tenth sheaf.

Compare this situation with, say, that of farmers whose staple crops are tubers such as potatoes or cassava/manioc. Such crops ripen in a year but may be safely left in the ground for an additional year or two. They can be dug up as needed and the remainder stored where they grew, underground. If an army or tax collectors want your tubers, they will have to dig them up tuber by tuber, as the farmer does, and then they will have a cartload of potatoes which is far less valuable (either calorically or at the market) than a cartload of wheat, and is also more likely to spoil quickly.[23] Frederick the Great of Prussia, when he ordered his subjects to plant potatoes, understood that, as planters of tubers, they could not be so easily dispersed by opposing armies.[24]

The "aboveground" simultaneous ripening of cereal grains has the inestimable advantage of being legible and assessable by the state tax collectors. These characteristics are what make wheat, barley, rice, millet, and maize the premier *political* crops. A tax assessor typically classifies fields in terms of soil quality and, knowing the average yield of a particular grain from such soil, is able to estimate a tax. If a year-to-year adjustment is required, fields can be surveyed and crop cuttings taken from a representative patch just before harvest to arrive at an estimated yield for that particular crop year. As we shall see, state officials tried to raise crop yields and taxes in kind by mandating techniques of cultivation; in Mesopotamia this included insisting on repeated ploughing to break up the large clods of earth and repeated harrowing for better rooting and nutrient delivery. The point is that with cereal grains and soil preparation, the planting, the condition of the crop, and the ultimate yield were more visible and assessable. Compare this, for example, with the attempt to assess and tax the commercial activity of buyers and sellers in the market. One reason for the official distrust and stigmatization of the merchant class in China was the simple fact that its wealth, unlike that of the rice planter, was illegible, concealable, and fugitive. One might tax a market, or collect tolls on a road or river junction where goods and transactions were more transparent, but taxing merchants was a tax collector's nightmare.

For purposes of measuring, dividing, and assessing, the simple fact that the cereal harvest consists ultimately of small grains, husked or unhusked, has enormous administrative advantages. Like grains of sugar or sand, cereal grains are almost infinitely divisible, down to smaller and smaller fractions

Figure 10. Beveled-rim (ration?) bowls.
Photo courtesy of Susan Pollock

and precisely measurable by weight and volume for accounting purposes. Units of grain served as standards of measurement and value for trade and tribute against which the value of
other commodities was calculated—including labor. The daily
food ration of the lowest class of laborers in Umma, Mesopotamia, was almost exactly two liters of barley measured out
in the beveled bowls that are among the most ubiquitous archaeological finds.

But why is there not a chickpea or lentil state? After all,
these are nutritious crops that can be grown intensively, and
their harvest consists of small seeds that can be dried, keep
well, and can as easily be divided and measured out in small
quantities as rations as the cereal grains. Here the decisive advantage of the cereal grains is their determinate growth and
hence virtually simultaneous ripening. The problem with most
of the legumes, from a tax collector's perspective, is that they
produce fruit continuously over an extended period. They

can be, and are, picked right along as they ripen—like beans or peas. If the tax collector arrives early, much of the crop will not yet have ripened, and if he arrives late, the taxpayer will probably have eaten, hidden, or sold much of the yield. One-stop shopping on the part of the tax collector works best for determinate-ripening crops. The cereal crops of the Old World were, in this sense, preadapted for state making. The New World—save for the mixed case of maize, which can be picked right along or left to mature and dry in the field—has few if any determinate, whole-field, simultaneously ripening crops, hence none of the harvest festival tradition that so dominates the Old World agricultural calendar. It leaves one to speculate whether determinate ripening was selected for by early Neolithic cultivators and if so, why, say, determinate ripening of chickpeas and lentils could not have been similarly selected for.

Even so, grain taxation is not foolproof. Though a given cereal crop, once planted, ripens simultaneously, the seasonality often allows for varying planting dates, so different fields may mature at slightly different times. It is also not uncommon for a tax-avoiding cultivator to harvest surreptitiously some of the grains before they are fully ripe in order to escape the tax. Archaic states endeavored, whenever possible, to mandate a planting time for a given district. In the case of irrigated wet rice, all adjoining fields are flooded at roughly the same time, and this alone dictates the (trans)planting schedule, not to mention the fact that rice is the only crop that will grow under these conditions.

Cereal grains also lend themselves well to bulk transport. Even under archaic conditions a cartload of grain could be drawn at a profit greater distances than almost any other food commodity. And where water transport was available, large quantities of grain could be shipped considerable distances, thereby greatly expanding the agricultural heartland an early state might hope to dominate and from which it could extract taxes. One account of the Third Dynasty of Ur (Ur III late third millennium BCE) claims that barges carried fully half of the entire barley harvest of the Ur region to royal depots.[25] Again, for the tax collector of early Mesopotamia and, for that matter, until the nineteenth century, the combination of an agrarian state and a navigable river or coastline was a marriage made in heaven. Rome, for example, found it cheaper to ship grain (usually from Egypt) and wine across the Mediterranean than to ship it overland by cart more than one hundred miles.[26]

Grain, because it has higher value per unit volume and weight than almost any other foodstuff, and because it stores comparatively well, was an ideal tax and subsistence crop. It could be left unhusked until it was needed. It was ideal for distributing to laborers and slaves, for requiring as tribute, for provisioning soldiers and garrisons, for relieving a food shortage or famine, or for feeding a city while resisting a siege. It is hard to imagine the early state without grain as a basis for its sinew and muscle.

Where grain, and therefore agrarian taxes, stopped, there too did the state's power begin to degrade. The power of the early Chinese states was confined to the arable drainage basins of the Yellow and Yangzi Rivers. Beyond this ecological and political heartland of fixed-field and irrigated rice farming lay the hard-to-tax, mobile pastoralists, hunter-gatherers, and shifting cultivators. They were defined as "raw" barbarians, who had "not yet entered the map." The territory of the Roman Empire, for all its imperial ambitions, did not extend much beyond the grain line. Roman rule north of the Alps was concentrated in what archaeologists term, after the Swiss site at which its artifacts were first found, La Tène zone, where population was denser, agricultural production more robust and towns (*oppida* culture) larger; outside this zone lay "Jastorf Europe," thinly populated and characterized by pastoralism and swiddening.[27]

This contrast is a salutary reminder that outside the earliest grain state lay most of the world and its population as well. The grain states were restricted to a narrow ecological niche that favored intensive agriculture. Beyond their horizon were a variety of what might be called nonappropriable subsistence practices, the most important of which were hunting and gathering, maritime fishing and collecting, horticulture, shifting cultivation, and specialized pastoralism.

Looked at from the perspective of a state tax collector, such forms of subsistence were fiscally sterile; they could not repay the cost of controlling them. Hunters and gatherers and maritime foragers were so dispersed and mobile, and their "takings" so diverse and perishable, that tracking them, let alone taxing them, was well-nigh impossible. Horticulturalists, who may well have domesticated roots and tubers well before grain was first planted, could hide a small plot in the forest and leave much of their harvest in the ground until they needed it. Swidden cultivators often planted some grain, but a typical swidden contained dozens and dozens of cultivars of differing maturity. Moreover, swiddeners moved their fields every few years and, occasionally, their dwellings as well. Specialized pastoralism, seen as an outgrowth of agriculture, confronts the would-be tax collector with a similar problem of dispersal and mobility. The Ottoman Empire, founded by pastoralists, found it exceptionally difficult to tax herders. They tried taxing them at the one moment of the year when they stopped to attend to lambing and shearing, but even this was logistically difficult. As Rudi Lindner, a student of Ottoman rule, concluded, "The Ottoman dream of a sedentary paradise with its predictable revenue from pacific farmers had no place for pastoral nomads." "The nomads followed small scale changes in climate to maximize their access to good pasture and sweet water; consequently they were always on the move."[28]

In one way or another, nongrain peoples—that it to say most of the world—embodied forms of livelihood and social organization that defeated taxation: physical mobility, dispersal, variable group and community size, diverse and invisible subsistence goods, and few fixed-point resources. It was not as if they were worlds apart, however. Quite to the contrary, as we have noted, exchange and trade flowed vigorously between them. The exchange, however, was uncoerced and depended on bartering and trading desirable goods from one ecological zone to another to mutual advantage. Those practicing a particular form of subsistence often came to be seen as a different kind of people, despite trading partnerships. To Romans, for example, a key defining characteristic of barbarians was that they ate dairy products and meat and not, as Romans did, grain. To the Mesopotamians, the "barbarian" Amorites were beyond the pale because they purportedly "know not grain . . . eat uncooked meat and do not bury their dead."[29]

The various forms of subsistence described above should not be seen as self-contained, impermeable categories. Groups can and did move between subsistence practices and often concocted hybrid practices that defied easy categorization. Nor should we discount the possibility that the choice of subsistence practices was often a political choice—a decision about positionality vis-à-vis the state.

WALLS MAKE STATES:

PROTECTION AND CONFINEMENT

Most towns in the Mesopotamian alluvium were, by the middle of the third millennium BCE, walled. The state, for the first time, had grown a defensive carapace. Although the sites were generally modest—anywhere from ten to thirty-three hectares on average—building and maintaining such a defensive perimeter, though it might be erected piecemeal, was labor intensive. A wall, in the crudest sense, tells us that there is something valuable being protected or held away from those outside. The existence of walls was an infallible proxy for the presence of permanent cultivation and food storage. And, as if to further confirm the association, when such a city-state collapsed and its walls were permanently breached, permanent cultivation was also likely to disappear from the area. It was common practice for a conquering city to tear down the walls of the town it had defeated. The existence of concentrated, valuable, lootable, fixed-point resources created, self-evidently, a powerful incentive to defend them. Their spatial concentration made it easier to protect them, and their value made the effort worthwhile. There is every reason why a peasantry would do what it could to hold on to its fields and orchards, its homes and its granaries, and its livestock as a matter of life and death. No wonder, then, that the Epic of Gilgamesh, a founding king, erects the city walls to protect his people. On that premise alone, might one see the creation of the state as a joint creation—a social contract, perhaps?— between cultivating subjects and their ruler (and his warriors and engineers) to defend their harvests, families, and livestock from attacks by other statelets or nonstate raider?

But the matter is more complicated. Just as a farmer may have to defend his crops against human and nonhuman predators, so state elites have an overwhelming interest in safeguarding the sinews of their own power: a cultivating population and its grain stores, its privileges and wealth, and its political and ritual powers. As Owen Lattimore and others have observed for the Great Wall(s) of China: they were built quite as much to keep Chinese taxpaying cultivators inside as to keep the barbarians (nomads) outside. City walls were thus intended to keep the essentials of state preservation inside. The so-called anti-Amorite walls between the Tigris and Euphrates may also have been designed more to keep cultivators in the state "zone" than to keep out the Amorites (who were, in any case, already settled in substantial numbers in the alluvium). The walls were, in the view of one scholar, a result of the vastly increased centralization of Ur III and were erected either to contain mobile populations fleeing state control or to defend against those who had been forcibly expelled. It was, in any event, "intended to define the limits of political control."[30] The control and confinement of populations as the reason and function of city walls depends in large part on demonstrating that the flight of subjects was a real preoccupation of the early state—the subject of Chapter 5.

WRITING MAKES STATES:
RECORD KEEPING AND LEGIBILITY

> To be governed is to be at every operation, at every
> transaction, noted, registered, counted, taxed, stamped,
> measured, numbered, assessed, licensed, authorized,
> admonished, prevented, reformed, corrected, punished.
> —Pierre-Joseph Prudhon

Peasantries with long experience of on-the-ground statecraft have always understood that the state is a recording, registering, and measuring machine. So when a government surveyor arrives with a plane table, or census takers come with their clipboards and questionnaires to register households, the subjects understand that trouble in the form of conscription, forced labor, land seizures, head taxes, or new taxes on croplands cannot be far behind. They understand implicitly that behind the coercive machinery lie piles of paperwork: lists, documents, tax rolls, population registers, regulations, requisitions, orders—paperwork that is for the most part mystifying and beyond their ken. The firm identification in their minds between paper documents and the source of their oppressions has meant that the first act of many peasant rebellions has been to burn down the local records office where these documents are housed. Grasping the fact that the state *saw* its land and subjects through record keeping, the peasantry implicitly assumed that *blinding* the state might end their woes. As an ancient Sumerian saying aptly puts it: "You can have a king and you can have a lord, but the man to fear is the tax collector." [31]

Southern Mesopotamia was the heartland of not one but several related state-making experiments between roughly 3,300 and 2,350 BCE. Like China's Warring States period or the later Greek city-states, the southern alluvium was the site of rivalrous city-polities whose fortunes waxed and waned. Among the best known were Kish, Ur, and, above all, Uruk. Something utterly remarkable and without historical parallel was taking place here. On one hand, groups of priests, strong men, and local chiefs were scaling up and institutionalizing structures of power that had previously used only the idioms of kinship. They were creating for the first time something along the lines of what we would call a state, though they could not possibly have understood it in those terms. On the other hand, thousands of cultivators, artisans, traders, and laborers were being, as it were, repurposed as subjects and, to this end, counted, taxed, conscripted, put to work, and subordinated to a new form of control.

It is at roughly this time that writing makes its first appearance. [32] The coincidence of the pristine state and pristine writing tempts one to the crude functionalist conclusion that would-be state makers invented the forms of notation that were essential to statecraft. But it would not be too strong to assert that it is virtually impossible to conceive of even the earliest states without a systematic technology of numerical record keeping, even if it took the Inka form of strings of knots (*quipu*). The first condition of state appropriation

(for whatever purpose) must be an inventory of available resources—population, land, crop yields, livestock, storehouse
stocks. This information is, however, like a cadastral survey,
a snapshot soon out of date. As appropriation proceeds, continuous record keeping is required—of grain deliveries, corvée labor performed, requisitions, receipts, and so on. Once a
polity comprises even a few thousand subjects, some form of
notation and documentation beyond memory and oral tradition is required.

A powerful case for linking state administration and writing is that it seems to have been used in Mesopotamia essentially for bookkeeping purposes for more than half a millennium before it even began to reflect the civilizational glories
we associate with writing: literature, mythology, praise hymns,
kings lists and genealogies, chronicles, and religious texts.[33]
The magnificent Epic of Gilgamesh, for example, dates from
Ur's Third Dynasty (circa 2,100 BCE), a full millennium after
cuneiform had been first used for state and commercial purposes.

What can one infer from the trove of cuneiform tablets
that have been recovered and translated about actual governance on the ground in Sumer? They reveal, at a minimum,
the massive effort through a system of notation to make a society, its manpower, and its production legible to its rulers
and temple officials, and to extract grain and labor from it.
Surely we know enough about even quite modern bureau-
cracies to realize that there is no necessary relation between
the records on the one hand and the facts on the ground on
the other. Documents are forged and fiddled for private advantage or to please superiors. Rules and regulations laid out

meticulously in the documents may be a dead letter on the ground. Land records may be corrupt, absent, or simply inaccurate. The order of the records office, like the order of the parade ground, too often masks rampant disorder in actual administration and on the battlefield. What the records can tell us, however, is something about the utopian, Linnaean order in statecraft that is implicit in the logic of record keeping, its categories, its units of measurement, and, above all, in the things it pays attention to. The "gleam in the eye" of what I think of as the "quartermaster state"—is most instructive. As a mark of this aspiration, the very symbol of kingship in Sumer was the "rod and line," almost certainly the tools of the surveyor.[34] We can see this state imagination at work in a brief examination of Mesopotamia and early Chinese administrative practice.

The earliest administrative tablets from Uruk (Level IV), circa 3,300–3,100 BCE, are lists, lists, and lists—mostly of grain, manpower, and taxes. The topics of the surviving tablets in order of frequency are barley (as rations and taxes), war captives, male and female slaves.[35] A preoccupation at Uruk IV and later in other centers is the population roll. As in all ancient kingdoms, maximizing population was an obsession that usually superseded the conquest of territory per se. Population—as producers, soldiers, and slaves—represented the wealth of the state. The city of Umma, a dependency of Ur, where a huge trove of tablets has been found dating from about 2,255 BCE, was especially precocious, occupying one hundred hectares and having between ten thousand and twenty thousand inhabitants—a large population to administer. At the core of Umma's project of legibility was a census

Figure 11. Cuneiform tablet depicting storehouse supplies
and withdrawals. Photo courtesy of the British Museum

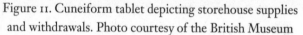

of population by location, age, and gender as the basis for as-
signing the head tax and corvée labor, and for conscription.
It was the "immanent" project, never realized in practice except perhaps for the temple economy and dependent labor
force. Landholdings, apparently both temple and private,
were designated by their size, the quality of their soil, and the
expected crop yield, which served as the basis for a tax assessment. Some of the Sumerian polities, especially Ur III, look
like command-and-control economies, heavily centralized
(on paper—or, rather, on tablet), militarized, and regimented,
resembling what we know of militarized Sparta among the
Greek city-states. One tablet records 840 rations of barley,

meted out, in all probability in the (mass produced?) beveled bowls holding one liter of barley. Other rations mention beer, groats, and flour. Labor gangs, whether of war captives, slaves, or corvée laborers, seem ubiquitous.

The entire exercise in early state formation is one of standardization and abstraction required to deal with units of labor, grain, land, and rations. Essential to that standardization is the very invention of a standard nomenclature, through writing, of all the essential categories—receipts, work orders, labor dues, and so on. The creation and imposition of a written code throughout the city-state replaced vernacular judgments and was itself a distance-demolishing technology that held sway throughout the small realm. Labor standards were developed for such tasks as ploughing, harrowing, or sowing. Something like "work points" were created, showing credits and debits in work assignments. Standards of classification and quality were specified for fish, oil, and textiles—which were differentiated by weight and mesh. Livestock, slaves, and laborers were classified by gender and age. In embryonic form, the vital statistics of an appropriating state aiming to extract as much value as possible from its land and people is already in evidence. How formidable this regimentation looked on the ground is another matter.

Writing appears in early China more than a millennium later along the Yellow River. It may have begun in the Erlitou cultural area, though no evidence survives. It is most famously known in the Shang Dynasty (1,600–1,050 BCE), through the finds of oracle bones used for divination. From then and on through the Warring States period (476–221 BCE), it was

continuously in use, particularly for purposes of state administration. Only with the famous, reforming, and short-lived
Qin Dynasty (221–206 BCE), however, does the nexus between writing and state making become clearest. The Qin,
rather like Ur III, was a systematizing, order-obsessed regime
that laid out a rather comprehensive vision of the total mobilization of its resources. On paper, at least, it was even more
ambitious. Neither in China nor in Mesopotamia was writing
originally devised as a means of representing speech.

A precondition of the standardization and simplification
the Qin aimed at was a reformed and unified script that eliminated a quarter of the ideograms, made it more rectilinear, and
applied it throughout its territory. Since the script was not a
transcription of a speech dialect, it had, inherently, a kind of
universality.[36] As with other early precocious states, the process of standardization was applied to coinage and to units of
weight, distance, and volume for, among other things, grain
and land. The intention was to eliminate a host of local, vernacular, and idiosyncratic practices of measurement so that,
for the first time, the ruler at the center could have a clear
view of the wealth, production, and manpower resources at
his disposal. It aimed at creating a centralized state rather
than merely a strong city-state that was content to extract
occasional tribute from a constellation of quasi-independent
satellite towns. Sima Qian, a court historian under the Han,
looked back favorably on Qin Emperor Shang Yang's accom-
plishment in fashioning his kingdom into an austere war machine: "For the fields, he opened up the *qian* and the *ma* (hori-
zontal and vertical pathways), and set up boundaries." "He

equalized the military levies and land tax and standardized the measures of capacity, weights and length."[37] Later, work norms and tools were standardized as well.

In the context of regional military rivalry with competing statelets, it was important to squeeze as much as possible from the realm. This meant creating and updating as complete an inventory of resources as possible, given the available techniques. Meticulous household registration to facilitate the head tax and conscription was a sign of power, as was a large and growing population. Captives were settled near the court, and regulations restricted population movement. One of the hallmarks of early statecraft in agrarian kingdoms was to hold the population in place and prevent any unauthorized movement. Physical mobility and dispersal are the bane of the tax man.

Land, happily for the tax collector, does not move. But as the Qin recognized private landholding, it conducted an elaborate cadastral survey connecting each piece of cropland with an owner/taxpayer. Land was classified by soil quality, crops sown, and variation in rainfall, which allowed tax officials to compute an expected yield and arrive at a tax rate. The Qin tax system also provided for estimates of standing crops on an annual basis, permitting, at least in theory, for tax adjustments according to actual harvests.

We have thus far concentrated on the intention of state officials, through writing, statistics, censuses, and measurement, to move beyond sheer plunder and to more rationally extract labor and foodstuffs from their subjects. This project, while perhaps the most important, is hardly the only policy by which a state attempts to sculpt the landscape of the polity to

make it richer, more legible, and more amenable to appropriation. Though the early state did not invent irrigation and water
control, it did extend irrigation and canals to facilitate transport and enlarge grain lands. Whenever it could it increased
both the numbers and legibility of its productive population by forced resettlement of subjects and war captives. The
"equal field" concept of the Qin was in large part to make sure
that all subjects had enough land to pay taxes and to provide
a population base for conscription. Under the Qin, reflecting the importance of population, the state not only forbade
flight but instituted a pro-natalist policy, with tax breaks to
women and their families who gave birth to new subjects. The
late-Neolithic resettlement camp was the kernel of the earliest
states, but much of early statecraft was an artful political landscaping to facilitate appropriation: more grain land, a larger
and more concentrated population, and the information soft-
ware made possible by written records that could make it all
more accessible to the state. Efforts at root and branch political landscaping may have been the undoing of the most ambitious early states. The superregimented Third Dynasty of Ur
lasted barely a century and the Qin only fifteen years.

If early writing is so inextricably bound to state making,
what happens when the state disappears? What little evidence
we do have suggests that without the structure of officials, administrative records, and hierarchical communication, literacy shrinks greatly if it does not disappear altogether. This
should not be surprising inasmuch as in the earliest states,
scriptural literacy was confined to a very thin veneer of the
population, most of whom were officials. From roughly 1,200
to 800 BCE, Greek city-states disintegrated in an era known

as the Dark Age. When literacy reappeared it no longer took the old form of Linear B but was an entirely new script borrowed from the Phoenicians. It was not as if all Greek culture disappeared in the interim. Instead, it took oral forms, and we owe both the *Odyssey* and the *Iliad*, later transcribed, to this period. Even the fragmentation of the Roman Empire, with its more extensive literary tradition, in the fifth century CE led to the near disappearance of literacy in Latin outside a few religious establishments. One suspects that in the earliest states, writing developed first as a technique of statecraft and was therefore as fragile and evanescent an achievement as the state itself.

What if we were to think of literacy in the earliest societies as one technology of communication, just as crop planting is one among many techniques of subsistence? The techniques of planting were known long before they found widespread use, and then only in particular ecological and demographic circumstances. In the same sense, it is not as if the world were "dark" until writing was invented, after which all societies adopted or aspired to adopt literacy. The first writing was, as well, an artifact of state building, concentration of population, and scale. It was inapplicable in other settings. One student of early writing in Mesopotamia suggested, admittedly speculatively, that writing was elsewhere resisted because of its indelible association with the state and taxes, just as ploughing was long resisted because of its indelible association with drudgery.

> [Why did] every distinctive community on the periphery reject the use of writing with so many archaeological cultures exposed to the complexity of southern Mesopotamia? One

could argue that this rejection of complexity was a conscious act. What is the reason for it? . . . Perhaps, far from being less intellectually qualified to deal with complexity, the peripheral peoples were smart enough to avoid its oppressive command structures for at least another 500 years, when it was imposed upon them by military conquest. . . . In every instance the periphery initially rejected the adoption of complexity even after direct exposure to it . . . and, in doing so, avoided the cage of the state for another half millennium.[38]

CHAPTER FIVE

Population Control:
Bondage and War

In the multitude of people is the king's honor, but in the want
of people is the destruction of the prince.
—Proverbs 14:28

If the multitudes scatter and cannot be retained, the city state
will become a mound of ruins.
—Early Chinese Manual of Governance

It is true, I admit, that [the Siamese kingdom] is of greater
extent than mine, but you must admit that the king of
Golconda rules over men, while the king of Siam rules over
forests and mosquitoes.
—King of Golconda to a Siamese visitor, circa 1680

In a large house with many servants, the doors may be left
open; in a small house with few servants, the doors must be
shut.
—Siamese saying

THE excess of epigraphs above is meant to signal the degree to
which concern over the acquisition and control of population
was at the very center of early statecraft. Control over a fertile

and well-watered patch of alluvium meant nothing unless it was made productive by a population of cultivators who would work it. To see the early states as "population machines" is not far off the mark, so long as we appreciate that the "machine" was in bad repair and often broke down, and not only because of failures in statecraft. The state remained as focused on the number and productivity of its "domesticated" subjects as a shepherd might husband his flock or a farmer tend his crops.

The imperative of collecting people, settling them close to the core of power, holding them there, and having them produce a surplus in excess of their own needs animates much of early statecraft.[1] Where there was no preexisting settled population that could serve as the nucleus of state forma-tion, a population had to be assembled for the purpose. This was the guiding principle of Spanish colonialism in the New World, the Philippines, and elsewhere. The *reducciones* or con-centrated settlements (often forced) of native peoples around a center from which Spanish power radiated were seen as part of a civilizing project, but they also served the nontrivial pur-pose of serving and feeding the conquistadores. Christian mission stations—of whatever denomination—among dis-persed populations begin in the same fashion, assembling a productive population around the station, from which con-version efforts radiated.

The means by which a population is assembled and then made to produce a surplus is less important in this context than the fact that it does produce a surplus available to non-producing elites. Such a surplus does not exist until the em-bryonic state creates it. Better put, until the state extracts and appropriates this surplus, any dormant additional productivity that might exist is "consumed" in leisure and cultural elaboration. Before the creation of more centralized political structures like the state, what Marshall Sahlins has described as the domestic mode of production prevailed.[2] Access to resources—land, pasture, hunting—was open to all by virtue of membership in a group, whether tribe, band, lineage, or family, that controlled those resources. Short of being cast out, an individual could not be denied direct and independent access to whatever means of subsistence the group in question disposed of. And in the absence of either compulsion or the chance of capitalist accumulation, there was no incentive to produce beyond the locally prevailing standards of subsistence and comfort. Beyond sufficiency in this respect, that is, there was no reason to increase the drudgery of agricultural production. The logic of this variant of peasant economy was worked out in convincing empirical detail by A. V. Chayanov, who, among other things, showed that when a family had more working members than nonworking dependents, it reduced its overall work effort once sufficiency was assured.[3]

The important point for our purpose is that a peasantry—assuming that it has enough to meet its basic needs—will not automatically produce a surplus that elites might appropriate, but must be compelled to produce it. Under the demographic conditions of early state formation, when the means of traditional production were still plentiful and not monopolized, only through one form or another of unfree, coerced labor—corvée labor, forced delivery of grain or other products, debt bondage, serfdom, communal bondage and tribute, and various forms of slavery—was a surplus brought into being. Each of the earliest states deployed its own unique mix of coerced labor, as we shall see, but it required a delicate balance between maximizing the state surplus on the one hand and the risk of provoking the mass flight of subjects on the other, especially where there was an open frontier. Only much later, when the world was, as it were, fully occupied and the means of production privately owned or controlled by state elites, could the control of the means of production (land) alone suffice, without institutions of bondage, to call forth a surplus. So long as there are other subsistence options, as Ester Boserup noted in her classic work, "it is impossible to prevent the members of the lower class from finding other means of subsistence unless they are made personally unfree. When population becomes so dense that land can be controlled it becomes unnecessary to keep the lower classes in bondage; it is sufficient to deprive the working class of the right to be independent cultivators"— foragers, hunter-gatherers, swiddeners, pastoralists.[4]

In the case of the earliest states, making the lower classes reliably unfree meant holding them in the grain core and preventing them from fleeing to avoid drudgery and/or bondage itself.[5] Do what it might to discourage and punish flight—and the earliest legal codes are filled with such injunctions—the archaic state lacked the means to prevent a certain degree of leakage under normal circumstances. In hard times occasioned by, say, a crop failure, unusually heavy taxes, or war, this leakage might quickly become a hemorrhage. Short of stemming the flow, most archaic states sought to replace their losses by various means, including wars to capture slaves, purchases of slaves from slave takers, and forced resettlement of whole communities near the grain core.

The total population of a grain state, assuming it controlled sufficient fertile land, was a reliable, if not infallible, indication of its relative wealth and military prowess. Aside from an advantageous position on trade routes and waterways or particularly clever rulers, agricultural techniques as well as the technology of warfare were both relatively static and depended largely on manpower. The state with the most people was generally richest and usually prevailed militarily over smaller rivals. One indication of this fundamental fact was that the prize of war was more often captives than territory, which meant that the losers' lives, particularly those of women and children, were spared. Many centuries later Thucydides acknowledges the logic of manpower by praising the Spartan general Brasidas for negotiating peaceful surrenders, thereby increasing the Spartan tax and manpower base at no cost in Spartan lives.[6]

Warfare in the Mesopotamian alluvium beginning in the late Uruk Period (3,500–3,100 BCE) and for the next two millennia was likewise not about the conquest of territory but rather about the assembling of populations at the state's grain core. Thanks to the original and meticulous work of Seth Richardson, we know that the vast majority of the wars in the alluvium were not those between the larger and well-known urban polities but, rather, the petty wars by each of those polities to conquer the smaller independent communities in its own hinterland to augment its laboring population and hence its power.[7] Polities aimed to assemble "unpacified," "scattered" people and to "herd non-state clients into state orders by both force and persuasion." This process, Richardson notes, is a continuing imperative inasmuch as states are simultaneously losing "their own constituent populations from and to non-state units." Though the state might presume to a fine-grained administration of its subjects, it was, in fact, in a constant struggle to compensate for the losses from flight and mortality by a largely coercive campaign to corral new subjects from among hitherto "untaxed and unregulated" populations. The Old Babylonian legal codes are preoccupied with escapees and runaways and the effort to return them to their designated work and residence.

THE STATE AND SLAVERY

Slavery was not invented by the state. Various forms of enslavement, individual and communal, were widely practiced among nonstate peoples. For pre-Columbian Latin America, Fernando Santos-Granaros has abundantly documented the many forms of communal servitude practiced, many of which persisted along with colonial servitude after the conquest.[8] Slavery, though generally tempered with assimilation and upward mobility, was common among manpower-hungry Native American peoples. Human bondage was undoubtedly known in the ancient Middle East before the appearance of the first state. As with sedentism and the domestication of grain that also predated state formation, the early state elaborated and scaled up the institution of slavery as an essential means to maximize its productive population and the surplus it could appropriate.

It would be almost impossible to exaggerate the centrality of bondage, in one form or another, in the development of the state until very recently. As Adam Hochschild observed, as late as 1800 roughly three-quarters of the world's population could be said to be living in bondage.⁹ In Southeast Asia all early states were slave states and slaving states; the most valuable cargo of Malay traders in insular Southeast Asia were, until the late nineteenth century, slaves. Old people among the so-called aboriginal people (*orang asli*) of the Malay Peninsula and hill peoples in northern Thailand can recall their parents' and grandparents' stories about much-dreaded slave raids.¹⁰

Provided that we keep in mind the various forms bondage can take over time, one is tempted to assert: "No slavery, no state." Moses Finley famously asked, "Was Greek Civilization based on Slave Labour?" and answered with a resounding and well-documented yes.¹¹ Slaves represented a clear majority— perhaps as much as two-thirds—of Athenian society, and the institution was taken completely for granted; the issue of abolition never arose. As Aristotle held, some peoples, owing to a lack of rational faculties, are, by nature, slaves and are best used, as draft animals are, as tools. In Sparta, slaves represented an even larger portion of the population. The difference, to which we shall return later, was that while most slaves in Athens were war captives from non-Greek-speaking peoples, Sparta's slaves were largely "helots," indigenous cultivators conquered in place by Sparta and made to work and produce communally for "free" Spartans. In this model the appropriation of an existing, sedentary grain complex by militarized state builders is far more explicit.

Imperial Rome, a polity on a scale rivaled only by its easternmost contemporary, Han Dynasty China, turned much of the Mediterranean basin into a massive slave emporium. Every Roman military campaign was shadowed by slave merchants and ordinary soldiers who expected to become rich

by selling or ransoming the captives they had taken personally. By one estimate, the Gallic Wars yielded nearly a million
new slaves, while, in Augustan Rome and Italy, slaves represented from one-quarter to one-third of the population. The
ubiquity of slaves as a commodity was reflected in the fact
that in the classical world a "standardized" slave became a unit
of measurement: in Athens at one point—the market fluctu-
ated—a pair of working mules was worth three slaves.

SLAVERY AND BONDAGE IN MESOPOTAMIA

In the earlier, less documented, and smaller city polities of
Mesopotamia the existence of slavery and other forms of
bondage is beyond question. Finley assures us, "The pre-
Greek world—the world of the Sumerians, Babylonians,
Egyptians, and Assyrians . . .—was, in a very profound sense,
a world without free men, in the sense in which the west has
come to understand the concept."[12] What is very much in
question, however, is the extent of slavery per se, the forms it
took, and how central it was to the functioning of the polity.[13]
The general consensus has been that while slavery was undoubtedly present, it was a relatively minor component of the
overall economy.[14] On the basis of my reading of the admittedly scarce evidence, I would dispute this consensus. Slavery,
while hardly as massively central as in classical Athens, Sparta,
or Rome, was crucial for three reasons: it provided the labor
for the most important export trade good, textiles; it supplied
a disposable proletariat for the most onerous work (for example, canal digging, wall building); and it was both a token
of and a reward for elite status. The case for the importance

of slavery in the Mesopotamian polities is, I hope to show, convincing. When other forms of unfree labor, such as debt bondage, forced resettlement, and corvée labor, are taken into account, the importance of coerced labor for the maintenance and expansion of the grain-labor module at the core of the state is hard to deny.

Part of the controversy over the centrality of slavery in ancient Sumer is a matter of terminology. Opinions differ in part because there are so many terms that could mean "slave" but could also mean "servant," "subordinate," "underling," or "bondsman." Nevertheless, scattered instances of purchase and sale of people—chattel slavery—are well attested, though we do not know how common they were.

The most unambiguous category of slaves was the captured prisoner of war. Given the constant need for labor, most wars were wars of capture, in which success was measured by the number and quality of captives—men, women, and children—taken. Of the many sources of dependent labor identified by I. J. Gelb—household-born slaves, debt slaves, slaves purchased on the market from their abductors, conquered peoples brought back and forcibly settled as a group, and prisoners of war—the last two appear to be the most significant.[15] Both categories represent the booty of war. On one list of 167 prisoners of war there appeared very few Sumerian or Akkadian (that is, indigenous) names; the vast majority had been taken from the mountains and from areas to the east of the Tigris River. One ideogram for "slave" in third-millennium Mesopotamia was the combination of the sign for "mountain" with the sign for "woman," signifying women taken in the course of military forays into the hills or perhaps bartered by slave takers in exchange for trade goods. The related ideogram "man" or "woman" joined to "foreign land" is also thought to refer to slaves. If the purpose of war was largely the acquisition of captives, then it makes more sense to see such military expeditions more in the light of slave raids than as conventional warfare.

The only substantial, documented slave institution in Uruk appears to have been the state-supervised workshops producing textiles that engaged as many as nine thousand women. They are described as slaves in most sources but also may have included debtors, the indigent, foundlings, and widows—perhaps like the workhouses of Victorian England. Several historians of the period claim that both women and juveniles taken as prisoners of war, complemented by the wives and children of debtors, formed the core of the textile workforce. Analysts of this large textile "industry" stress how critical it was to the position of elites, who were dependent for their power on a steady flow of metals (copper in particular) and other raw materials from outside the resource-poor alluvium. This state enterprise provided the key trade good that could be exchanged for these necessities. The workshops represented a sequestered "gulag" of captive labor that supported a new strata of religious, civil, and military elites. Nor was it insignificant demographically. Various estimates put the Uruk population at around forty thousand to forty-five thousand in the year 3,000 BCE. Nine thousand textile workers alone would represent at least 20 percent of Uruk's inhabitants, not counting the other prisoners of war and slaves in other sectors of the economy. Providing grain rations for these workers and other state-dependent laborers required a formidable apparatus of assessment, collection, and storage.[16]

Other Uruk documents refer frequently to unfree workers and particularly to female slaves of foreign origin. They were, according to Guillermo Algaze, a primary source of workers at the disposal of the Uruk state administration.[17] The scribal summaries of laboring groups (both foreign and native) employ the identical age and sex categories as those used to describe "state-controlled herds of domestic animals." "It would appear, therefore, that in the minds of the Uruk scribes and in the eyes of the institutions that employed them, such laborers were conceptualized as 'domesticated' humans, wholly equivalent to domestic animals in status."[18]

What else can we say about the organization, work, and treatment of prisoners and slaves? An exceptional and quite detailed picture—despite fragmentary sources—is afforded by a close examination of 469 slaves and prisoners of war brought to Uruk and held in a "house of prisoners" during the reign of Rim-Anum (c. 1,805 BCE).[19] "It is most likely that houses of prisoners existed elsewhere in Mesopotamia and in other areas of the ancient Middle East."[20] The "house" functioned as something of a labor-supply bureau. The captives represented a wide spectrum of skills and experience and were disbursed to individuals, temples, and military officers as boatmen, gardeners, harvest workers, herdsmen, cooks, entertainers, animal tenders, weavers, potters, craftspeople, brewers, road menders, grinders of grain, and so on. The house—not apparently a workhouse itself—received flour in return for the labor it provided. Care was taken to farm out small labor crews and to relocate them frequently to minimize the danger of revolt or escape.

Other evidence about slaves and prisoners of war indicates that they were not well treated. Many are shown in neck fetters or being physically subdued. "On cylinder seals we meet frequent variants of a scene in which the ruler supervises his men as they beat shackled prisoners with clubs."[21] There are many reports of captives being deliberately blinded, but it is impossible to know how common the practice was. Perhaps the strongest evidence of brutal treatment is the general conclusion by scholars that the servile population did not reproduce itself. In lists of prisoners, it is striking how many are listed as dead—whether from the forced march back or from overwork and malnutrition is not clear.[22] Why valuable manpower would be so carelessly destroyed is, I believe, less likely to be owing to a cultural contempt for war captives than to the fact that new prisoners of war were plentiful and relatively easy to acquire.

The strongest circumstantial evidence for slaves and captive prisoners comes, as one might expect, from later periods after Ur III, when cuneiform texts are more abundant. Whether one can make a case for reading such evidence back to Ur III or find it applicable to our understanding of the Uruk period (c. 3,000 BCE) is highly questionable. In these later periods, much of the apparatus of slave "management" is evident. There are bounty hunters whose specialty it is to locate and return runaway slaves. The escapees are subdivided into "recent" escapees, those long-gone, "deceased" escapees,

Figure 12. Prisoners in neck fetters. Photo courtesy
of the Iraq Museum, Baghdad, Dr. Ahmed Kamel

and "returned" escapees, though it seems as if few of the runaway slaves were ever recaptured.[23] Throughout these sources
there are accounts of populations fleeing a city for causes as
varied as hunger, oppression, epidemics, and warfare. Many
captive prisoners of war are undoubtedly among them, though
it is unknown whether they fled back to their place of origin,
or to another town, which would surely have welcomed them,
or to pastoralism. In any event, absconding was a preoccupation of alluvium politics; the later well-known code of Ham-
murabi fairly bristles with punishments for aiding or abetting
the escape of slaves.

A curious confirmation of the conditions of slave and enslaved debtors in Ur III comes from reading a utopian hymn
"against the grain." Prior to the construction of a major temple

Figure 13. The grinding room in early–second millennium palace at Ebla. Reprinted from Postgate, *Early Mesopotamia: Society and Economy at the Dawn of History*

(Eninnu) there was a ritual suspension of "ordinary" social relations in favor of a radical egalitarian moment. A poetic text describes what does *not* happen in this ritual of exception:

> The slave woman was an equal of her mistress
> The slave walked at his master's side
> The orphan was not delivered to the rich one
> The widow was not delivered to the powerful one
> The creditor did not enter one's house
> He [the ruler] undid the tongue of the whip and the goad
> The master did not strike the slave on the head
> The mistress did not slap the face of the slave women
> He canceled the debts[24]

The depiction of a utopian space, by negating the ordinary woes of the poor, weak, and enslaved, provides a handy portrait of quotidian conditions.

EGYPT AND CHINA

Whether slavery existed at all in ancient Egypt—at least in the Old Kingdom (2,686–2,181 BCE)—is hotly debated. I am in no position to settle the matter, which, in any case, depends on what one considers "slavery" and what period of ancient Egypt we are describing.[25] The issue may be, as one recent commentator describes it, a distinction without a difference, inasmuch as corvée and work quotas for subjects were so onerous. An admonition to become a scribe captures the burdens of subjects: "Be a scribe. It saves you from toil and protects you from all manner of work. It spares you from bearing hoe and mattock, so that you do not carry a basket. It sunders you from plying the oar and spares you torment, as you are not under many lords and numerous masters."[26]

Wars of capture on the Mesopotamian model were conducted during the Fourth Dynasty (2,613–2,494 BCE), and "foreign" prisoners of war were branded and forcibly resettled on royal "plantations" or within other temple and state institutions where the labor quotas were demanding. From what I can gather, though the scale of early slavery is uncertain, it seems clear that during the Middle Kingdom period (2,155–1,650 BCE) something very close to chattel slavery existed on a large scale. Captives were brought back from military campaigns and both owned and sold by slave merchants. "The demand for shackles was so great that the temples regularly placed orders for their manufacture."[27] Slaves seem to have been passed on by inheritance inasmuch as inventories of inherited property listed livestock and people. Debt bondage was also common. Later, under the New Kingdom (sixteenth to eleventh century BCE), the large-scale military campaigns in the Levant and against the so-called sea peoples generated thousands of captives, many of whom were taken back to Egypt and resettled en masse as cultivators or as laborers in often fatal quarries and mines. Some of these captives were probably among the royal tomb builders who staged one of the first recorded strikes against palace officials who had failed to deliver their rations. "We are in extreme destitution . . . lacking in every staple. . . . Truly we are already dying, we are no longer alive" wrote a scribe on their behalf.[28] Other conquered groups were required to produce annual tribute in metal, glass, and, it seems, slaves as well. What is in doubt for the Old and Middle Kingdoms is not, I think, the existence of something very like slavery, but rather its overall importance to Egyptian statecraft.

What we know of the brief Qin Dynasty and the early Han following it reinforce the impression that the earliest states are population machines seeking to maximize their manpower base by all possible means.[29] Slavery was just one of those means. The Qin lived up fully to its reputation as an early effort at total and systematic rule. It had markets for slaves in the same way as it had markets for horses and cattle. In areas outside dynastic control, bandits seized whomever they could and sold them at slave markets or ransomed them. The capital of both dynasties was filled with war captives seized by the state, by generals, and by individual soldiers. As with most early warfare, military campaigns were mixed with "privateering," in which the most valuable loot comprised the number of captives who could be sold. It seems that much of the cultivation under the Qin was carried out by captive slaves, debt slaves, and "criminals" condemned to penal servitude.[30]

The major technique for assembling as many subjects as possible, however, was the forced resettlement of the entire population—but especially women and children—of conquered territories. The captives' ritual center was destroyed, and a replica rebuilt at Xinyang, the Qin capital, signifying a new symbolic center. As was also typical for early statecraft in Asia and elsewhere, the prowess and charisma of a leader was indexed by his capacity to assemble multitudes around his court.

SLAVERY AS "HUMAN RESOURCES" STRATEGY

Finally, war helped to a great discovery—that men as well as animals can be domesticated. Instead of killing a defeated enemy, he might be enslaved; in return for his life he might

be made to work. This discovery has been compared in
importance to that of the taming of animals. . . . By early
historic times slavery was a foundation of ancient industry
and a potent instrument in the accumulation of capital.
—V. Gordon Childe, *Man Makes Himself*

Adopting for the moment the purely strategic view of a
quartermaster in charge of manpower needs can help clarify
why slavery, in the form of war captives that it usually took,
had several advantages over other forms of surplus appropria-
tions. The most obvious advantage is that the conquerors take
for the most part captives of working age, raised at the expense of another society, and get to exploit their most productive years. In a good many cases the conquerors went out
of their way to seize captives with particular skills that might
be useful—boat builders, weavers, metal workers, armorers,
gold- and silversmiths, not to mention artists, dancers, and
musicians. Slave taking in this sense represented a kind of
raiding and looting of manpower and skills that the slaving
state did not have to develop on its own.[31]

Insofar as the captives are seized from scattered locations
and backgrounds and are separated from their families, as was
usually the case, they are socially demobilized or atomized
and therefore easier to control and absorb. If the war captives came from societies that were perceived in most respects
as alien to the captors, they were not seen as entitled to the
same social consideration. Having, unlike local subjects, few
if any local social ties, they were scarcely able to muster any
collective opposition. The principle of socially detached servants—Janissaries, eunuchs, court Jews—has long been seen

as a technique for rulers to surround themselves with skilled but politically neutralized staff. At a certain point, however, if the slave population is large, is concentrated, and has ethnic ties, this desired atomization no longer holds. The many slave rebellions in Greece and Rome are symptomatic, although Mesopotamia and Egypt (at least until the New Kingdom) appeared not to have slavery on this scale.

Women and children were particularly prized as slaves. Women were often taken into local households as wives, concubines, or servants, and children were likely to be quickly assimilated, though at an inferior status. Within a generation or two they and their progeny were likely to have been incorporated into the local society—perhaps with a new layer of recently captured slaves beneath them in the social order. If manpower-hungry polities like, say, Native American societies or Malay society historically are any indication, it is common to find pervasive slavery together with rapid cultural assimilation and social mobility. It was not uncommon, for example, for a male captive of the Malays to take a local wife and, in time, organize slave-taking expeditions of his own. Providing that slaves were constantly being acquired, such societies would remain slave societies, but, viewed over several generations, earlier captives would have become nearly indistinguishable from their captors.

Women captives were at least as important for their reproductive services as for their labor. Given the problems of infant and maternal mortality in the early state and the need of both the patriarchal family and the state for agrarian labor, women captives were a demographic dividend. Their reproduction may have played a major role in alleviating the otherwise unhealthy effects of concentration and the domus. Here I cannot resist the obvious parallel with the domestication of livestock, which requires taking control over their reproduction. The domesticated flock of sheep has many ewes and few rams, as that maximizes its reproductive potential. In the same sense, women slaves of reproductive age were prized in large part as breeders because of their contribution to the early state's manpower machine.

The continuous absorption of slaves at the bottom of the social order can also be seen to play a major role in the process of social stratification—a hallmark of the early state. As earlier captives and their progeny were incorporated into the society, the lower ranks were constantly replenished by new captives, further solidifying the line between "free" subjects and those in bondage, despite its permeability over time. One imagines, as well, that most of the slaves not put to hard labor were monopolized by the political elites of the early states. If the elite households of Greece or Rome are any indication, a large part of their claim to distinction was the impressive array of servants, cooks, artisans, dancers, musicians, and courtesans on display. It would be difficult to imagine the first elaborate social stratification in the earliest states without war-captive slaves at the bottom and elite embellishment, dependent on those slaves, at the top.

There were, of course, many male slaves outside the households. In the Greco-Roman world, captive enemy combatants—particularly if they had offered stiff resistance—might be executed, but many more were ransomed or brought back as war booty. A state that depends on a population of scarce producers is unlikely to squander the essential prize of early warfare. Though we know precious little about the disposition of male war captives in Mesopotamia, in the Greco-Roman territories they were deployed as a kind of disposable proletariat in the most brutal and dangerous work: silver and copper mining, stone quarrying, timber felling, and pulling oars in galleys. The numbers involved were enormous, but because they worked at the sites of the resources, they were a far less visible presence—and far less a threat to public order—than if they had been near the court center.[32] It would be no exaggeration at all to think of such work as an early gulag, featuring gang labor and high rates of mortality. Two aspects of this sector of slave labor deserve emphasis. First, mining, quarrying, and felling timber were absolutely central to the military and monumental needs of the state elites. These needs in the smaller Mesopotamian city-states were more modest but no less vital. Second, the luxury of having a disposable and replaceable proletariat is that it spared one's own subjects from the most degrading drudgery and thus forestalled the insurrectionary pressures that such labor well might provoke, while satisfying important military and monumental ambitions. In addition to quarrying, mining, and logging, which only desperate or highly paid men will undertake voluntarily, we might include carting, shepherding, brick making, canal digging and dredging, potting, charcoal making, and pulling oars on boats or ships. It is possible that the earliest Mesopotamian states traded for many of these commodities, thereby outsourcing the drudgery and labor control to others. Nevertheless, much of the materiality of state making depends centrally on such work, and it matters whether those doing it are slaves or subjects. As Bertolt Brecht, in his poem "Questions from a Worker Who Reads," asked:

> Who built the Thebes of the seven Gates?
> In the books you will read the names of kings.
> Did the kings haul up the lumps of rocks?
> And Babylon many times demolished,
> Who raised it up so many times?

BOOTY CAPITALISM AND STATE BUILDING

A sure sign of the manpower obsession of the early states, whether in the Fertile Crescent, Greece, or Southeast Asia, is how rarely their chronicles boast of having taken territory. One looks in vain for anything resembling the twentieth-century German call for *lebensraum*. Instead, the triumphal account of a successful campaign, after praising the valor of the generals and troops, is likely to aim at impressing the reader with the amount and value of the loot. Egypt's victory over Levantine kings at Kadesh (1,274 BCE) is not just a paean to the pharaoh's bravery but a record of the plunder, and in particular of the livestock and prisoners—so many horses, so many sheep, so many cattle, so many people.[33] The human prisoners are, here as elsewhere, often distinguished for their skills and crafts, and one imagines that something of an inventory was made of the talent the conquerors had acquired. The conquerors were on the lookout for generic manpower and, simultaneously, for the craftsmen and entertainers who would enhance the luster of the conquerors' courts. The towns and villages of the defeated peoples were generally destroyed so that there was nothing to go back to. In theory, the plunder belonged to the ruler, but in practice the loot was divided up, with the generals and individual soldiers taking their own livestock and prisoners to keep, ransom, or sell. Thucydides, in his history of the Peloponnesian Wars, has several accounts of such conquests and adds that most of the wars were fought when the grain was ripe, so that it too could be seized as plunder and fodder.[34]

Max Weber's concept of "booty capitalism" seems applicable to a great many such wars, whether conducted against competing states or against nonstate peoples on its periphery. "Booty capitalism" simply means, in the case of war, a military campaign the purpose of which is profit. In one form, a group of warlords might hatch a plan to invade another small realm, with both eyes fixed on the loot in, say, gold, silver, livestock, and prisoners to be seized. It was a "joint-stock company," the business of which was plunder. Depending on the soldiers, horses, and arms that each of the conspirators contributes to the enterprise, the prospective proceeds might be divided proportionally to each participant's investment. The enterprise is, of course, fraught, inasmuch as the plotters (unless they are merely financial backers) potentially risk their lives. To be sure, such wars may have other strategic aims, like the control of a trade route or the crushing of a rival, but for the early states, the taking of loot, particularly human captives, was not a mere by-product of war but a key objective.[35] Slaving wars were systematically conducted by many of the earliest states in the Mediterranean as a part of their manpower needs. In many cases—in early Southeast Asia and in imperial Rome—war was seen as a route to wealth and comfort. Everyone from the commanders down to the individual soldier expected to be rewarded with his share of the plunder. To the degree that men of military age were engaged in slaving expeditions, as they were in imperial Rome, it posed a problem for the labor force in grain and livestock production at home. In time, the huge influx of slaves allowed landowners—and peasant soldiers—to replace much of the agrarian labor force with slaves who were not themselves subject to conscription.

Despite the relative absence of hard evidence on the extent of slavery in Mesopotamia and early Egypt, one is tempted to speculate that the slave sector erected over the grain module in the early states was, even if of modest size, an essential component in the creation of a powerful state. The pulses of captive slaves alleviated many of the manpower needs of an otherwise demographically challenged state. Perhaps most crucial was the fact that slaves, a few skilled workers excepted, were concentrated in the most degrading and dangerous labor, often away from the domus, which was central to the material and symbolic sinews of its power. If such states had had to extract such labor exclusively from their own core subjects, they would have run a high risk of provoking flight or rebellion—or both.

THE PARTICULARITY OF MESOPOTAMIAN SLAVERY AND BONDAGE

Historians and archaeologists are fond of saying, as we have noted, that "the absence of evidence is not evidence of absence." The evidence of slavery and bondage we have examined is hardly absent, but it is sparse enough to have convinced a number of scholars that slavery and bondage were insignificant. In what follows, I hope to suggest the reasons why slavery should seem less obtrusive and central in the Mesopotamian evidence than in Greece or Rome. Those reasons have to do with the modest size and geographical reach of the Mesopotamian polities, the origins of their slave population, the possible "subcontracting" of unfree labor, the importance of corvée labor from the subject population, and the potential role of communal forms of bondage. In the course of examining the scholarship on labor in Mesopotamia, I find that in the case of at least some monumental building projects, the labor required of the subject (not slave) population may have been less than often supposed, and that it may even have been accompanied by ritual feasting on the completion of the monument.[36]

Three obvious reasons why Third Millennium Mesopotamia might seem less of a slave-holding society than Athens or Rome are the smaller populations of the earlier polities, the comparably scarce documentation they left behind, and their relatively small geographic reach. Athens and Rome were formidable naval powers that imported slaves from throughout the known world, drawing virtually all their slave populations far and wide from non-Greek and non-Latin speaking societies. This social and cultural fact provided much of the foundation for the standard association of state peoples with civilization on the one hand and nonstate peoples with barbarism on the other. Mesopotamian city-states, by contrast, took their captives from much closer to home. For that reason, the captives were more likely to have been more culturally aligned with their captors. On this assumption, they might have, if allowed, more quickly assimilated to the culture and mores of their masters and mistresses. In the case of young women and children, often the most prized captives, intermarriage or concubinage may well have served to obscure their social origins within a couple of generations.

The origins of prisoners of war is a further complicating factor. Most of the literature on slavery in Mesopotamia concerns prisoners of war who spoke neither Akkadian nor Sumerian. Yet it is evident that intercity warfare in the alluvium was common. If, in fact, a significant portion of the captives came from intercity warfare for one another's subjects, and from hitherto independent local communities, then, given their shared culture, it is plausible that the captives would have become ordinary subjects of their captor's city-state without much further ado—perhaps without even being formally enslaved. The greater the cultural and linguistic differences between slaves and their masters, the easier it is to draw and enforce the social and juridical separation that makes for the sharp demarcation typical of slave societies.

In Athens in the fifth century BCE, for example, there was a substantial class, more than 10 percent of the population, of *metics*, usually translated as "resident aliens." They were free to live and trade in Athens and had the obligations of citizenship (taxes and conscription, for example) without its privileges. Among them were a substantial number of ex-slaves. One must surely wonder whether the Mesopotamian city-states met a substantial portion of their insatiable labor needs by absorbing captives or refugees from culturally similar populations. In this case such captives or refugees would probably appear not as slaves but as a special category of "subject" and perhaps would be, in time, wholly assimilated.

Just as most Western consumers never directly experience the conditions under which the material foundations of their lives are reproduced, so for the Greeks at Athens, that roughly half of the slave population working in the quarries, mines, forests, and galleys was largely invisible. On a far more modest scale the early Mesopotamian states had need of a male labor force to quarry stone, mine copper for armaments, and provide timber for construction, firewood, and charcoal. As these activities would have been carried out at a substantial distance from the floodplain, it would have been relatively invisible to subjects at the center, though not to state elites. The phenomenon known as "the Uruk Expansion"—the discovery of Uruk cultural artifacts in the hinterlands and in the Zagros Mountains—represents, it seems, a foray to create or guard trade routes for vital goods not available in the alluvium.[37] Though it is certain that slaves were seized in this expansion area, it is unclear whether Uruk directly used slaves and war captives in this primary extraction or whether it exacted tribute in these materials from subjugated communities—or, for that matter, traded grain, cloth, and luxury goods for them. In any case, such coerced labor would have taken place at arm's length from Uruk—subcontracted perhaps to trading partners—and might therefore leave few if any cuneiform traces.

Finally, there are two forms of communal bondage that were widely practiced in many early states and that bear more than a family resemblance to slavery but are unlikely to appear in the textual record as what we think of as slavery. The first of these might be called mass deportation coupled with communal forced settlement. Our best descriptions of the practice come from the neo-Assyrian Empire (911–609 BCE), where it was employed on a massive scale. Although the neo-Assyrian Empire falls much later than our main temporal focus, some scholars claim that such forms of bondage were used much earlier in Mesopotamia, Egypt's Middle Kingdom, and the Hittite Empire.[38]

Mass deportation and forced settlement was, in the neo-Assyrian Empire, systematically applied to conquered areas. The entire population and livestock of the conquered land were marched from the territory at the periphery of the kingdom to a location closer to the core, where they were forcibly resettled, the people usually as cultivators. Although, as in other slaving wars, some captives were "privately" appropriated and others formed into labor gangs, what was distinctive about deportation and forced settlement was that the bulk of the captive community was kept intact and moved to a site where its production could be more easily monitored and appropriated. Here, the manpower and grain–centralizing machine is at work but at a wholesale level, taking entire agrarian communities as modules and placing them at the service of the state. Even allowing for the exaggerations of the scribes, the scale of the population transfers was unprecedented. More than 200,000 Babylonians, for example, were moved to the core of the neo-Assyrian Empire, and the total deportations appear staggering.[39] There were specialists in deportations. Officials conducted elaborate inventories of the captured populations—their possessions, their skills, their livestock—and were charged with provisioning them en route to their new location with a minimum of losses. In some cases, it seems that the captives were resettled on land abandoned earlier by other subjects, implying that forced mass resettlement may have been part of an effort to compensate for mass exoduses or epidemics. Many of the captives were referred to as "saknutu," which means "a captive made to settle the soil."

The neo-Assyrian policy is not historically novel. Though we have no idea whether it was common in Mesopotamia, it has been the practice of conquest regimes throughout history—in Southeast Asia and the New World in particular. For our purpose, however, what is most important is that these resettled populations would not necessarily have appeared in the historical record as slaves at all. Once resettled, especially if they were not markedly different culturally, they might well have become ordinary subjects, scarcely distinguishable over time from other agrarian subjects. Some of the confusion over whether earlier Sumerian terms (for example, *erin*) should be translated as "subject," as "prisoner of war," as "military colonist," or simply as "peasant" may well derive from the various classes of subjects that reflect the origins of their "subjecthood."

A final genre of bondage that is historically common and also might not appear in the historical record as slavery is the model of the Spartan helot. The helots were agricultural communities in Laconia and Messinia dominated by Sparta. How they came to be so dominated is a matter of dispute. Messinia seems to have been conquered in war, but some claim that the helots were either those who chose not to participate in warfare or who were collectively punished for an earlier revolt. They were, in any case, distinguished from slaves. They remained in situ as whole communities, were annually humiliated in Spartan rituals, and like the subjects of all archaic agrarian states were required to deliver grain, oil, and wine to their masters. Aside from the fact that they had not been forcibly resettled as war deportees, they were in all other respects the enserfed agricultural servants of a thoroughly militarized society.

Here, then, is another archaic formula by which the necessary manpower-and-grain complex was assembled that could serve as the surplus-yielding module of state building. It is conceivable, but quite unknowable, that some of the Mesopotamian city-states originated in the conquest or displacement of an agrarian population in situ by an external military elite. In this context, Nissen cautions us to heavily discount the rhetoric stigmatizing nonstate peoples and urges us to recall the constant interchange between mountains and lowlands. He claims, "Even the massive settlement of the Mesopotamian plain of the middle of the fourth millennium may have been part of this process."

"Tempted by the written record we have . . . internalized the viewpoint of the lowland inhabitants."[40] The fact that the place names Ur, Uruk, and Eridu are not Sumerian in origin hints at the possibility of an incursion—or the seizure of control by the militarized faction of an existing agrarian society. It is also conceivable that the grain core was expanded and replenished by the forced resettlement of war captives from the hinterland and from other cities. In either of these cases, such early societies would not have appeared superficially to be slave societies. And in fact, they would not have been slave societies in quite the Athenian or Roman sense. Yet the central role of bondage and coercion in creating and maintaining the grain-and-manpower nexus of the early agrarian state would be perfectly evident.

A SPECULATIVE NOTE ON DOMESTICATION, DRUDGERY, AND SLAVERY

States, we know, did not invent slavery and human bondage; they could be found in innumerable prestate societies. What states surely did invent, however, are large-scale societies based systematically on coerced, captive human labor. Even when the proportion of slaves was far less than in Athens, Sparta, Rome, or the neo-Assyrian Empire, the role of captive labor and slavery was so vital and strategic to the maintenance of state power that it is difficult to imagine these states persisting long without it.

What if we were, as a fruitful conjecture, to take seriously Aristotle's claim that a slave is a tool for work and, as such, to be considered as a domestic animal as an ox might be? After all, Aristotle was serious. What if we were to examine slavery, agrarian war captives, helots, and the like as state projects to domesticate a class of human servitors—by force—much as our Neolithic ancestors had domesticated sheep and cattle? The project, of course, was never quite realized, but to see things from this angle is not entirely far-fetched. Alexis de Tocqueville reached for this analogy when he considered Europe's growing world hegemony: "We should almost say that the European is to the other races what man himself is to the lower animals; he makes them subservient to his use, and when he cannot subdue, he destroys."[41]

If we substitute for "Europeans" "early states," and for "other races" "war captives," we do not greatly distort the project, I think. The captives, individually and collectively, became an integral part of the state's means of production and reproduction, a part, if you will, along with the livestock and grain fields of the state's own domus.

Pushed even farther, I believe the analogy has an illuminating power. Take the question of reproduction. At the very center of domestication is the assertion of human control over the plant's or animal's reproduction, which entails confinement and a concern for selective breeding and rates of reproduction. In wars for captives, the strong preference for women of reproductive age reflects an interest at least as much in their reproductive services as in their labor. It would be instructive, but alas impossible, to know, in the light of the epidemiological challenges of early state centers, the importance of slave women's reproduction to the demographic stability and growth of the state. The domestication of non-slave women in the early grain state may also be seen in the same light. A combination of property in land, the patriarchal family, the division of labor within the domus, and the state's overriding interest in maximizing its population has the effect of domesticating women's reproduction in general.

The domesticated plough animal or beast of burden lifts much of the drudgery from man's back. Much the same could obviously be said for slaves. Over and above the drudgery of plough agriculture, the military, ceremonial, and urban needs of the new state centers required forms of labor in terms of both kind and scale that had no precedent. Quarrying, mining, galley oaring, road building, logging, canal digging, and other menial tasks may have been, even in more contemporary times, the sort of work performed by convicts, indentured laborers, or a desperate proletariat. It's the sort of work away from the domus that "free" men—including peasants—shun. Yet such dangerous and heavy work was necessary to the very survival of the earliest states. If one's own agrarian population could not be made to do this work without risking desertion or rebellion, then a captive, domesticated, alien population must be made to do it. That population could be acquired only by slavery—the long-standing, ultimately unsuccessful, and last attempt to realize Aristotle's vision of the human tool.

CHAPTER SIX

Fragility of the Early State: Collapse as Disassembly

HE more one reads about the earliest states, the greater one's astonishment at the feats of statecraft and improvisation that brought them into being in the first place. Their vulnerability and fragility were so manifest that it is their rare appearance and even rarer persistence that requires explanation. The image conjured by early state building is that of the four- or five-tiered human pyramid attempted by schoolchildren. It usually collapses before it is completed. When, against the odds, it is built to the apex, the audience holds its breath as it sways and trembles, anticipating its inevitable collapse. If the tumblers are lucky, the last one, representing its peak, has a fleeting moment to pose in triumph for the spectators. To pursue the metaphor a bit farther, the individual segments of the pyramid are, taken singly, quite stable; we might call them the elementary units or building blocks. The elaborate struc-

ture they create, however, is wobbly and ramshackle. That it soon falls apart is hardly surprising; what's remarkable is that it was done at all.

As a political structure assembled atop a settled farming community, the state shared the general vulnerabilities of sedentary grain communities in general. Sedentism was, as we have noted earlier, not a once-and-for-all achievement. Over the roughly five millennia of sporadic sedentism before states (seven millennia if we include preagriculture sedentism in Japan and the Ukraine), archaeologists have recorded hundreds of locations that were settled, then abandoned, perhaps resettled, and then again abandoned. The reasons for abandonment and reoccupation typically remain obscure. Possible contributing factors include climate change, resource depletion, disease, warfare, and migration to areas of greater abundance. The general recession of whatever modest fixed settlements existed before 10,500 BCE was almost surely due to the Younger Dryas cold snap—"the big freeze." Another sudden and widespread demise around 6,000 BCE of a cultural complex associated with settlement, documented for the Jordan Valley and known as the Prepottery Neolithic Phase B (PPNB), has been variously attributed to climate change, disease, soil depletion, shrinking water sources, and demographic pressure. The key point is that, as a subspecies of sedentary grain communities, states were subject to the same perils of dissolution as sedentary communities in general, as well as to the fragility particular to states as political entities.

Consensus about the fragility of the first archaic states seems unanimous; about the causes of this fragility there is no consensus, and what little evidence we have is rarely dispositive. Robert Adams, whose knowledge of the early Meso-potamian states is unsurpassed, expresses some astonishment at the Third Dynasty of Ur (Ur III), in which five kings suc-ceeded one another over a hundred-year period. Though it too collapsed afterward, it represented something of a rec-ord of stability as compared with the dizzying comings and goings of other kingdoms. Adams discerns a cycle of cen-tralization of resources followed by an irregular but irrevers-ible decline, which he associates with a push for decentral-ization and "local self-sufficiency."[1] Norman Yoffee, Patricia McAnany, and George Cowgill, who have reexamined, far more than others, the very concept of "collapse," believe that "concentrations of power in early civilizations were typically fragile and short-lived."[2] Cyprian Broodbank, who has sur-veyed Mesopotamian, Levantine, and Mediterranean polities more generally, reaches the same conclusion, pointing to the "bewildering pattern of foundation, abandonment, expansion and shrinkage, as local or wider opportunities and adversity dictated."[3]

What might "collapse" mean, anyway—as in the phrases "the collapse of Ur III," around 2,000 BCE; "the collapse of the Old Kingdom Egypt," around 2,100 BCE; "the collapse of the Minoan Palatial Regime" on Crete, around 1,450 BCE? At the very least it means the abandonment and/or destruction of the monumental court center. This is usually interpreted not merely as a redistribution of population but as a substan-tial, not to say catastrophic, loss of social complexity. If the population remains, it is likely to have dispersed to smaller settlements and villages.[4] Higher-order elites disappear; monumental building activity ceases; use of literacy for administrative and religious purposes is likely to evaporate; larger-scale trade and redistribution is sharply reduced; and specialist craft production for elite consumption and trade is diminished or absent. Taken together, such changes are often understood to be a deplorable regression away from a more civilized culture. In this respect, it is just as essential to emphasize what such events do *not* necessarily mean. They do not necessarily mean a decline in regional population. They do not necessarily mean a decline in human health, well-being, or nutrition, and, as we shall see, may represent an improvement. Finally, a "collapse" at the center is less likely to mean a dissolution of a culture than its reformulation and decentralization.

The history of the term "collapse" and the melancholic associations it evokes are worth reflecting on. Our initial knowledge of and wonder at the archaic state come from what might be called the heroic period of archaeology, around the turn of the twentieth century, when the monumental centers of early civilizations were being pinpointed and excavated. Apart from a justified awe at the cultural, aesthetic, and architectural achievements of these early civilizations, there was something of a competitive imperial scramble to appropriate both their lineage of grandeur and their artifacts. Finally, through the schoolbooks and the museums, the prevailing standard images of these early states have become icons: the pyramids and mummies of Egypt, the Athenian Parthenon, Angkor Wat, the warrior tombs at Xian. So when these archaeological superstars evaporated, it seemed as if it were the end of an entire world. What in fact was lost were the beloved objects of classical archaeology: the concentrated ruins of the relatively rare centralized kingdoms, along with their written record and luxuries. To revert briefly to the human pyramid metaphor, it was as if the apex of the assemblage, the part on which all attention was riveted, had suddenly vanished.

When the apex disappears, one is particularly grateful for the increasingly large fraction of archaeologists whose attention was focused not on the apex but on the base and its constituent units. Their cumulative knowledge of shifting settlement patterns, structures of trade and exchange, rainfall, soil structure, and changing mixes of livelihood strategies allows us to see a great deal more than the apparently gravity-defying apex. From their findings we are able not only to discern some of the probable causes of "collapse" but, more important, to interrogate just what collapse might mean in any particular case. One of their key insights has been to see much that passes as collapse as, rather, a disassembly of larger but more fragile political units into their smaller and often more stable components. While "collapse" represents a reduction in social complexity, it is these smaller nuclei of power—a compact small settlement on the alluvium, for example—that are likely to persist far longer than the brief miracles of statecraft that lash them together into a substantial kingdom or empire. Yoffee and Cowgill have aptly borrowed from the administrative theorist Herbert Simon the term "modularity": a condition wherein the units of a larger aggregation are generally independent and detachable—in Simon's terms, "nearly decomposable."[5] In such cases the disappearance of the apical center need not imply much in the way of disorder, let alone trauma, for the more durable, self-sufficient elementary units. Echoing Yoffee and Cowgill, Hans Nissen cautions us against mistaking "the end of a period of centralization as a 'collapse' and regarding the phase during which a once unified area was split up into smaller parts as a politically troubled period."[6]

Neither sedentism nor state building, which depended utterly upon it, was a once-and-for-all achievement. There are periods—protracted ones—in which large aggregations of population disappeared and in which sedentism itself was reduced to a mere shadow of its former self. From roughly 1,800 until 700 BCE—more than a millennium—settlements in Mesopotamia covered less than a quarter of their previous area, and urban settlements were only one-sixteenth as frequent as during the previous millennium. The effect was regionwide, so it cannot be associated with purely local contingencies such a harsh ruler, a local war, or a particular crop failure. Such large-scale dispersals call for larger regionwide causes, such as climate variation, invasions and displacement by pastoralists, or major disruptions in trade, or for slower-acting but still regionwide environmental deterioration that might suddenly reach a critical threshold. There seems to be no consensus on which causes were most significant, but there is no doubt that ruralization rather than urbanization dominated Mesopotamia for more than a thousand years after the fall of Ur III, apparently owing to pastoralist incursions.[7]

Quite apart from a climatological deus ex machina such as the Younger Dryas, the two-to-four-century cold snap beginning 6,200 BCE, or the Little Ice Age—events that massively constrain what is ecologically possible—it is essential to acknowledge the fundamental structural vulnerability of the grain complex on which all early states rested. Sedentism arose in very special and circumscribed ecological niches, particularly in alluvial or loess soils. Later—much later—the first centralized states arose in even more circumscribed ecological settings where there was a large core of rich, well-watered soils and navigable waterways, capable of sustaining a good number of cereal-growing subjects. Outside these rare and favorable sites for state creation, foraging, hunting, and pastoral people continued to flourish.

State-making sites were above all structurally vulnerable to subsistence failures that had little to do with how adept or incompetent their rulers were. First and foremost of these structural vulnerabilities was the fact that they depended overwhelmingly on a single annual harvest of one or two cereal staples. If that harvest failed because of drought, flood, pests, storm damage, or crop diseases, the population was in mortal danger—as were their rulers who depended on the surplus they produced. These populations were also, as we have seen, in far greater danger from the infectious diseases that affected them and their livestock because of crowding than were dispersed foragers. And finally, as we shall explore, the reliance of elites on a surplus, together with the logic of transportation, meant that the state relied far more heavily on the population and resources located closest to the core, a reliance that could undermine its stability.

The earliest states were, then, delicate balancing acts; a lot had to go right for them to have anything but a brief life. In early Southeast Asia, for example, it was rare for a kingdom to last for more than two or three reigns—and any number of problems, not all of the kingdom's own making, could easily bring it down. The periodic demise of most kingdoms was "overdetermined," and because the difficulties they faced were so manifold, a coroner-archaeologist would be hard-pressed to single out a particular cause of death.

EARLY STATE MORBIDITY: ACUTE AND CHRONIC

The first pristine states in the Middle East, China, and the New World were operating in totally uncharted territory. There was no way that their founders and subjects could anticipate the ecological, political, and epidemiological perils that awaited them. Since the problems were without precedent, they were hard to fathom. Once in a while, especially when there are written sources, the reason for a state's demise is fairly clear: a successful invasion by another culture that replaces its enemy, for example, a destructive war between states, or a civil war or insurrection within the state. More commonly, however, the reasons behind the state's disappearance are more obscure and insidious, or else are catastrophic events, such as flood, drought, or crop failure, which may have deeper, cumulative causes. Such causes, I believe, are of particular interest to us for at least two reasons. First, unlike more contingent events like an invasion, they have a systematic character that may be linked directly to state processes. As such, they afford us a unique window on the structural contradictions of the ancient state. Second, such causes are likely to be slighted by most historical analyses, as they appear to have no direct, proximate human agent behind them and often leave no obvious archaeological signature behind to identify themselves. Evidence for their role in state mortality is speculative as well as circumstantial, but there is reason to believe their importance has been greatly underestimated.

Disease: Hypersedentism, Movement, and the State

We have explored at considerable length the rise of infectious diseases associated with crowding and the domestication of livestock. There is every reason to believe that the creation of states atop the Neolithic grain-and-animal complex would have greatly aggravated the exposure of early state populations to devastating epidemics. The reasons have to do with scale, trade, and warfare.

The towns that first emerged on the wetland fringes of the alluvium prior to states had, at their apogee, populations on the order of five thousand. The early states, by contrast, were typically four times larger and, occasionally, ten times as great. With the increase in the order of magnitude came an increase in the magnitude of risk. If the sudden eclipse of Phase B of the Prepottery Neolithic (PPNB) around 6,000 BCE was due, as some believe, to epidemic disease, the greater scale of the early states more than two millennia later would have made them that much more prone to epidemics. The larger populations would have represented a more substantial human and animal reservoir for infectious disease, and the effect of both crowding and numbers, on the geometric logic of transmission, would have spread it quickly.

Germs and parasites move with people and animals. While limited trade over some distance predated states, the volume and geographical reach of trade expanded exponentially with the rise of larger, expansive elites seeking to maximize their wealth and put it on display. States themselves required resources on a far grander scale than early sedentary communities, and resources of a different order. The result was an explosion of overland and, especially, waterborne trade. Students of early trade Guillermo Algaze and David Wengrow go so far as to refer to the "Uruk world system" around 3,500 to 3,200 BCE as an integrated world of trade and exchange stretching from the Caucasus in the north to the Persian Gulf in the south and from the Iranian Plateau in the east to the Eastern Mediterranean in the west.[8] Uruk and its competitors required resources from afar that were not available in the alluvium: copper and tin for tools, weapons, armor, and both decorative and utilitarian objects; timber and charcoal; limestone and quarried rock for building; silver, gold, and gems for sumptuary display. In exchange for these goods the statelets of the alluvium dispatched textiles, grain, pottery, and artisanal products to their trading partners. The effect, for our purposes, of this vast enlargement of the commercial sphere is that it similarly enlarged the sphere of transmitted diseases, bringing hitherto separate pools of diseases into contact for the first time. In this respect, the "Uruk world system," despite the grandiosity of the term, may well have prefigured, on a smaller scale, the integration of the Chinese, Indian, and Mediterranean disease pools around the year 1 BCE that is seen to have touched off the world's first devastating pandemics, such as the sixth-century CE Plague of Justinian, which killed between thirty million and fifty million people. Trade, responsible for much of the monumental splendor of the alluvium statelets, may, ironically, have played as large a role in their disappearance.

States are notorious for another activity: warfare, which has enormous epidemiological consequences. In terms of demography alone there is nothing like warfare for the mass movement and relocation of populations. An army or, for that matter, a mass of fleeing refugees or captives represents a moving module of infection, contracting and transmitting many of the diseases traditionally associated with war: cholera, typhus, dysentery, pneumonia, typhoid fever, and the like. The line of march of armies or refugees has long been known to mark a line of infection from which civilians seek, if they can, to flee. When, as in the case of ancient warfare, the major prize consists of captives who are marched back to the victor's kingdom, the consequences for infectious diseases are much the same as with trade, but perhaps on a larger scale. Among the captives, of course, were the enemy's four-footed livestock, which would have brought their own diseases and parasites along to the victor's capital.

How important were trade and warfare-borne diseases in the eclipse of early states? It's impossible to know for sure, as the archaeological record provides little in the way of evidence. My hunch is that they may have been responsible for a good many of the otherwise unexplained sudden abandonments of population centers in the ancient world. Working back from what we do know about epidemics in the Roman and medieval world may help make this hunch more plausible. As the diseases of crowding were novel, there was no way early populations could know the mechanisms by which they were spread. But the knowledge that outbreaks of lethal epidemics were associated with the shipping trade, overland caravans, armies, and their captives must have taken hold very early.[9]

The first instinct of a threatened townspeople would have been to isolate the first cases and wall off the town from any further contacts with the presumed sources of contamination. Quarantine and the isolation of maritime travelers (later institutionalized as *lazarretti*) must have arisen in one form or another along with new and dreaded epidemics. At the same time, even the earliest town dwellers must have understood that flight and dispersal from the site of a lethal epidemic represented the best hope of avoiding becoming infected. Their instinct was to scatter as quickly as possible to the countryside (where they were undoubtedly feared), and the earliest states would have been hard-pressed to stop them.

If this understanding of the response to early epidemics is broadly correct, then it provides a plausible scenario for disease-driven disappearance of major settlements. Once the epidemic was established, and assuming for the moment that the bulk of the population remained in the urban center, it might well kill enough of the population to destroy the city's viability as a state center. On the more realistic assumption that most of the population would have managed to flee, the result, though less lethal, would nevertheless have emptied out the urban center on which the state depended. Either scenario could, in short order, extinguish the state center as a node of power. The second scenario, however, need not entail a significant decline in the total population but rather its dispersal to safer, more rural locations. In one documented example, a devastating plague in the 1,320s BCE that came to Egypt from the Hittites sparked a famine, as surviving cultivators resisted taxes and often deserted their fields, while unpaid soldiers turned to banditry.[10] There is no way of knowing for certain how frequently epidemics brought down the earliest states, but, amplified by warfare, invasions, and trade, diseases were a prominent cause of deurbanization in late Imperial Rome and in medieval Europe. In 166 CE Roman troops returning from a campaign in Mesopotamia brought home an infectious disease that may have killed a quarter to a third of Rome's population.[11]

Ecocide: Deforestation and Salinization

That the first states were pristine creations deserves to be foregrounded in any analysis of their rise and demise. As earlier noted, there was no way that their subjects or elites could have foreseen that the unique assemblage of grain, people, and animals they presided over might have the epidemiological consequences they experienced. In a similar fashion, no one could have anticipated that the unprecedented burden of this assemblage would also generate unique and unsustainable demands on the surrounding environment. Of the environmental limits that were most likely to threaten the existence of the state, I examine two of the most important: deforestation and salinization.[12] Each is well documented in the ancient world from the earliest times. They differ, for the most part, from epidemic diseases in that they operate on a longer term; they are more gradual or, better put, more insidious than sudden. An epidemic, one imagines, was capable of devastating a city in a matter of weeks. A shortage of fuelwood or the gradual siltation of canals and rivers resulting from deforestation was more a matter of gradual economic suffocation—quite as lethal but far less spectacular.

The southern Mesopotamian alluvium was itself the natural erosive product of the Tigris and Euphrates, moving soil from the upper watershed and depositing it on the floodplain. Early agrarian societies depended, in this sense, on the dividend of nutrients transported downstream for millennia by the rivers. With the growth of large settlements, however, this process entered a new phase, as the need grew for timber and firewood not available in the wetlands of the alluvium. There is abundant evidence for the deforestation of the Euphrates upstream from Mari at the beginning of the third millennium BCE, owing to some combination of deforestation for timber and fuel with overgrazing.[13]

The early state's appetite for wood was nearly insatiable and far exceeded what even a sizable sedentary community might have required. In addition to clearing land for agriculture and grazing, and the need for wood for cooking and heating, house construction, and pottery kilns, the early state required huge quantities of wood for metallurgy, iron smelting, brick making, salt curing, mining supports, shipbuilding, monumental architecture, and lime-plaster—this last requiring huge amounts of fuelwood to prepare. Given the difficulties of transporting wood any appreciable distance, a state center would have very quickly have exhausted the modest supplies close to its core settlement. Located, as virtually all early states were, on a navigable waterway, usually a river, it could take advantage of the buoyancy of wood and the current of the river to cut timber on the banks upstream from the center.

The practicalities of logging and transportation again dictated that trees be felled as close to the river as possible

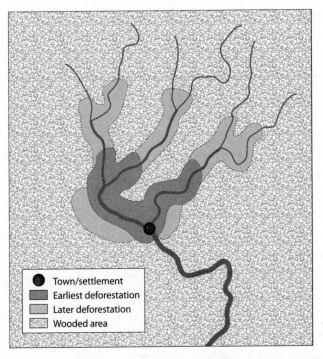

Figure 14. Pattern of upstream deforestation
from a hypothetical state center

to minimize labor. As the nearby upstream banks were de-
forested, the wood had to come from farther and farther upstream and/or from smaller trees that could be more easily
gotten to the bank, where they could be floated downstream.
There is abundant evidence for deforestation in the classical
world from the Athenian quest for naval timber in Macedonia
and the shortage of timber in the Roman Republic.[14] Much
earlier, by 6,300 BCE, in the Neolithic town of Ain Ghazal,

there were no more trees within walking distance of the settlement, and fuelwood had become scarce. As a result, the community dispersed into scattered hamlets, as did a good many other Jordan Valley Neolithic settlements when they exceeded the carrying capacity of their local woodlots.[15]

A nearly infallible sign that a city-state faces a shortage of easily available firewood close at hand is the proportion of its requirements that is supplied by charcoal. Although charcoal is essential for high-temperature applications such as firing pottery, lime slaking, and smelting, it is unlikely to be used for domestic purposes unless nearby firewood has been exhausted. The singular advantage of charcoal is that it contains far more heat value per unit weight and volume than raw wood and can therefore be transported greater distances economically. Its disadvantage, of course, is that it has to be burned twice and is far more wasteful of wood. The less local firewood within easy gathering distance, the more likely it will be replaced by charcoal from a distance.

A shortage of fuelwood may constrain the growth of a city-state, but deforestation of the watershed upstream from the city poses other, more serious problems. The first of these problems is erosion and siltation. While the earliest states were creatures of the alluvium and its silt, the pace of siltation from a watershed either stripped of vegetation or simply cleared for crops carried its own unique dangers of increased erosion that could not easily have been foreseen. Because the first states were based in very low-gradient alluvium, their waterways were slow-moving most of the year; this meant that the silt tended to settle out as the current slackened. If the city-state depended heavily on irrigation, its canals would tend to choke with silt—further slowing the flow—requiring, at the very least, corvée labor to dredge them lest the fields they served go out of production.

Another threat deforestation posed was catastrophic rather than insidious. Forests—in ancient Mesopotamia they included oak, beech, and pine especially—had the effect of holding the late winter rains and slowly releasing their moisture by percolation beginning in May. The effect of deforestation or agricultural clearance was that the watershed released the rains and the silt they carried far more quickly, making for a faster and more violent flood pulse.[16] This could have several effects that might threaten a city-state's viability. If, as often happens, the process of siltation has raised the river bed to the level near that of the surrounding land, the river becomes exceptionally erratic, jumping from one channel to another as each silts up. The gradual siltation coupled with an inundation and high water might touch off a major, catastrophic flood. Historically, China's Yellow River is the textbook example of massive floods and radically fluctuating paths to the sea, responsible for millions of deaths. Even Jericho, one of the largest prestate Neolithic settlements, appears to have succumbed to watershed damage in the middle of the ninth millennium BCE: "The enemy was flood water and mud flows," writes Steven Mithen. "Jericho was in perpetual danger as increased rainfall and vegetation clearance destabilized sediments on the Palestine hills that could then be carried to the edge of the village by nearby wadis."[17] Short of a catastrophic flood that might destroy much of a city-state and its crops, the river might also change course at flood tide, leaving an existing city high and dry, marooned from its major artery of transportation and commerce.

One last and more speculative consequence of deforestation and siltation is its role in the propagation of malaria. It has been suggested that malaria is a "disease of civilization," in the sense that it arose with land clearance for agriculture. J. R. McNeill intriguingly suggests that this may be related to deforestation and river morphology. A silt-bearing river crossing a low-gradient coastal plain will, as it slows, deposit more silt. As the silt accumulates, it creates its own levee or barrier, blocking its passage to the sea and causing it to back up and spread out laterally, creating malarial wetlands that are both anthropogenic and perhaps uninhabitable.[18]

Salinization and soil exhaustion are two further anthropogenic results of the grain-and-irrigation state that may come to threaten its existence. All irrigation water contains dissolved salts. As plants do not take it up, it accumulates over time in the soil and, unless leached out by flushing, will kill them. A short-term solution only, flushing raises the water table, and as the salt persists, the flushing eventually brings it closer to the surface, where it enters the plant roots. Barley is more tolerant of salt than wheat, so one adaptation to increasing salinization is to plant barley instead of the generally more desirable wheat. Even with barley, however, if the water table and hence the salts are nearer the surface, the yields are dramatically lowered.[19] The low gradient and low rainfall of southern Mesopotamia aggravate the problem, and Adams, the expert in these matters, is convinced that progressive salinity was the major factor in the ecological decline of the region after 2,400 BCE.[20] Mesopotamian farmers had to fallow their grain fields every second or third year so as to maintain a viable yield. Agricultural texts from the Ur III period refer to nearby fields as "located at brackish water," in "a salty place," on "salty soil," and containing "heaps of salt" in order to explain the low cereal yields.[21]

It is quite likely that even in the rich alluvium, where irrigation-induced salinization was not the major problem, yields of cereals over time declined. After all, there was little experience up to this time with the continuous annual cropping of the same plot of land. Ain Ghazal experienced declining yields even before the first states, and, given the intensity of cereal cultivation at the core of the grain states, one suspects that the average yields would have declined in much the same fashion. Pasture lands may have been overgrazed as well, lowering their livestock-carrying capacity.

In understanding the fragility of the early states and the cause of their disappearance, we might usefully distinguish cases of "sudden death" (for example, the disappearance of Larsa in 1,720 BCE) from those of debilitation and eventual demise. Epidemics and great floods, though they may arise from cumulative underlying effects, are examples of the former. States obliterated in this way go out like a light, though much of the population may survive by flight and dispersal. The cases of siltation, declining yields, and salinization may appear in the historical record as a steady or irregular dwindling—a drifting away of population—or more frequent crop failures. There would be in such cases not necessarily any dramatic turning point, but rather a nearly imperceptible fading away. "Collapse" is far too histrionic a term to apply to such processes. They may be so common as to represent, for the state subjects involved, a familiar routine of dispersal and rearrangement of settlement and subsistence routines. Only for state elites might it have been experienced as a tragedy of "collapse."

Politicide: Wars and Exploitation of the Core

That the issue of "collapse" should arise at all is essentially an artifact of the rise of walled settlements with monumental centers, and the common mistaken assumption that such central places are "civilization" itself. There are any number of occasions, as we have noted, when prestate sedentary communities are, for one reason or another, abandoned temporarily or permanently. Such events, noted by archaeologists, may involve substantial numbers of people, but they are unlikely to be "historical news" so long as the community is not a walled state center. The stones and rubble matter; they provide both an impressive site of excavation, museum artifacts, and often an iconic lineage for a nation's glorious past. Civilizations that, like Srivijaya on Sumatra, built with perishable materials and now are all but vanished hardly appear in the history book, while Angkor Wat and Borobudur live on as luminous centers.

The state no more invented war than it did slavery. It did, however, once again scale up these institutions as major state activities. This transformed what had been modest but constant prestate raids for captives into something like a war with other states for the same purposes. In a war for captives between two states the losing state was, virtually by definition, effaced. Voilà! "Collapse"! The usual practice was to kill or carry off most of the population, destroy the shrines, burn houses and crops: in short to obliterate the losing state altogether. The exception was peaceful capitulation by one party, often followed by tribute and occasionally the occupation of the defeated land by settlers brought by the victor—a gentler alternative that eliminated the original state no less. When the polities at war were many, of comparable size, and in the same neighborhood, as was the case for the Mesopotamian alluvium, the "Warring States" of pre-Qin China, the Greek city-states, and the Mayan states—so-called "peer polities"—statelets rose and fell in rapid succession. Collapse was commonplace.

The constant warfare and jockeying for manpower further contributed to the fragility of the early states. First, and most obvious, it diverted manpower resources to wall building, defensive works, and offensive operations that might otherwise have been employed in producing food for a population not much above the subsistence level. Second, it forced the founders and builders of a city-state to choose a site and layout where military defense considerations might prevail over material abundance. This may well have resulted in states that, while more easily defended, were economically more precarious.

Despite the potential mercenary rewards of warfare for the victors, there was of course the danger of death and captivity to consider. One imagines that many subjects of the peer polities did whatever they could to avoid conscription, including flight from the state. A state that appeared to be losing its war would find its manpower leaking away. (One thinks of the massive desertions of poor whites from the Confederacy in the last stages of the U.S. Civil War in 1864.) Thucydides writes of the Athenian coalition unraveling as the campaign against Syracuse was failing: "With the enemy on equal terms with us, our slaves were beginning to desert. As for the foreigners in our service, those who were conscripted are going back to their cities as quickly as they can."[22] As manpower was the lifeblood of these states, a decisive defeat could well presage the collapse of the state itself.[23]

Finally, the city-state might have as easily been destroyed by internal conflict: battles for succession, civil wars, and insurrections. What is perhaps distinctive about internal strife is that there was a new and valuable prize worth commanding: a walled, surplus-producing grain core, with its population, livestock, and stores. Struggles to control an advantageous location were never trivial, even among prestate societies, but the advent of the early states raised the stakes largely because they represented a stock of fixed capital—canals, defensive works, records, storehouses, and, often, a valuable location with respect to soil, water, and trade routes. These assets were nodes of power that were not surrendered lightly and, one imagines, provoked more ferocious, no-holds-barred struggles for local power.

Whether as a prize of interstate war or of civil conflict, the grain-population complex remained the nucleus of political power. In interstate war and raids by nonstate peoples, the victor either sought to destroy this complex and transfer its movable assets to its own core or, failing that, to make it a tributary core. In the case of internal war, the battle was for the monopoly rights to appropriate the resources that the core represented.

To understand why the early state may have often dug its own grave by overexploiting the core region around the court, it helps to return to the basic constraints of transportation and appropriation. As illustrated by the sharply rising costs of firewood and hence the growing domestic use of charcoal, overland appropriation of bulk commodities is exponentially more expensive and soon becomes prohibitive as distance increases. This logic essentially delineates the practical reach of the state so long as transportation technology remains static. Assuming draft animals and carts on a flat alluvial plain, the reach of the earliest states for grain requisitions is unlikely to have extended much beyond a radius of roughly forty-eight kilometers. The crucial exception, of course, is water-borne transport, which, thanks to the radical reduction of friction, greatly extends the state's catchment area for bulk commodities like grain. An agrarian core could then be defined as the zone from which bulk commodities can be brought to the center without transportation costs becoming prohibitively expensive. The key fact, however, is that the most lucrative zone of control is the area closest to capital or easily reachable by navigable water routes. It is therefore within this zone that one finds the symbols and resources of power: grain stores, major shrines, administrative staff, praetorian guards, central markets, the most productive, best-watered agricultural lands, and, not least, the abode of the palace and temple elites.

It was this core zone that was the key to state power and cohesion. It was also the state's Achilles' heel, as it was this zone that was likely to be squeezed first and hardest in any crisis.[24] Precisely because this zone was closest at hand, most valuable, and dense with resources, it would, in a pinch, yield the most manpower and grain. An audacious ruler, one with military or monumental ambitions, one threatened by invasion or by internal enemies, would be tempted, as the line of least resistance, to draw resources from this core. Two facts made this a very dangerous gamble—one that could bring down states. First, for an agrarian kingdom always liable to the vagaries of rainfall, weather, pests, and human and crop diseases, the annual yield, even in this most reliable of agrarian ecologies, was extremely variable. In ordinary circumstances the "yield" elites might extract from this zone would vary widely. If elites insisted on a steady, let alone expanding, take from this zone in terms of grain and labor—on insulating itself from the normal fluctuations in output—then the core agrarian population would bear the potentially ruinous brunt of harvest fluctuations despite its own tenuous subsistence. As in all agrarian economies, the key issue in class relations is which class absorbs the inevitable shocks of a bad year—or, in other words, which class ensures its economic security at the expense of whom.

A second factor to recall in the case of pristine states was the quite rudimentary knowledge the state had of the actual acreage planted, the likely and the actual yields, district by district, for wheat and barley. Though the state knew a lot more about the vital core than about outlying areas, it was quite likely to confiscate too much grain in a bad year, leaving its subjects on the edge of starvation. That is, quite apart from rapaciousness, the first states lacked the fine-grained knowledge that would have made it easier to modify their appropriation in line with the capacity of their subjects to pay. They were, as a colleague of mine once said, "all thumbs and no fingers for fine-tuning."[25] The results of their misjudgment were also compounded by the inability to monitor the rapaciousness of their own tax collectors on the ground, intent on appropriating for themselves.

In an emergency, when maximizing tax revenue was a matter of survival, pressing on the core region was well-nigh irresistible, even though it might risk provoking flight and/ or rebellion. Outlying areas were not a realistic alternative. They were likely to be more marginal agriculturally, with lower and more variable yields; the revenues that could be appropriated from them were partly nullified by transportation costs; and the knowledge of these resources and control over the administrative apparatus that might appropriate them diminished radically with distance from the center. An elite, believing itself in mortal danger or seized with celestial ambitions, would have had little compunction in adopting survival strategies that risked killing the goose that laid the golden egg: the grain core. What is read retrospectively as "collapse" may often, I speculate, have been triggered by resistance and flight by desperate subjects in the core in situations like this.

Students of what "collapse" might actually have meant for the Mesopotamian states in the third millennium BCE point to the same issue of who assumes the burden of risk: "Since it is unlikely that the central authority will cut its costs in proportion to the reduction in revenue obtained from some elements of the society, it is highly likely that the tax burden will be increased for the remainder."[26] Evidence from the later stages of the Akkadian Dynasty (circa 2,200 BCE) indicates that the core of the kingdom was periodically squeezed, as it was both the juiciest and closest source of revenue. Core officials could and did require more grain to be planted and fallows to be shortened in order to maximize immediate returns at the cost of long-run productivity. Two centuries later, when Ur was threatened, it appears, by Amorite incursions, the defending generals pressed so hard on Ur's cultivators for grain that they either resisted or fled. The collapse of the manpower-grain state is captured in this passage from the famous Lamentation over Ur: "Hunger filled the city like water . . . its king breathes heavily in his palace, *all alone*, its people dropped their weapons."[27]

Egypt in the late third millennium BCE, a far larger and more consolidated kingdom than Mesopotamia's twenty-odd contending peer polities, was also apparently a state pressing relentlessly on its core agrarian population for grain and labor, depressing living standards.[28] The fact that the fertile strip along the Nile was hemmed in by deserts on each side made it possible to press the population harder than would have been feasible with a peasantry with more running room. Some commentators stress the bare-bones "kit" of the cultivating subjects and sumptuary laws that excluded 90 percent of the population from wearing certain clothing, owning prestige goods, or celebrating certain rituals reserved for the elite.[29]

Lacking the sort of demographic data that might allow us to track population movements, it is, alas, impossible to discover whether the volume of flight from the core increased as more and more grain and labor was extracted from its population. Assuming that flight was possible and common, was a state, by acquiring war captives and forcibly resettling them at the core, able to compensate for any leakage—slow or fast—of the hard-pressed subjects fleeing that core?

PRAISING COLLAPSE

Why deplore "collapse," when the situation it depicts is most often the disaggregation of a complex, fragile, and typically oppressive state into smaller, decentralized fragments?[30] One simple and not entirely superficial reason why collapse is deplored is that it deprives all those scholars and professionals whose mission it has been to document ancient civilizations of the raw materials they require. There are fewer important digs for archaeologists, fewer records and texts for historians, and fewer trinkets—large and small—to fill museum exhibits. There are splendid and instructive documentaries on archaic Greece, Old Kingdom Egypt, and mid-third millennium Uruk, but one will search in vain for a portrayal of the obscure periods that followed them: the "Dark Age" of Greece, the "First Intermediate Period" of Egypt, and the decline of Uruk under the Akkadian Empire. Yet there is a strong case to be made that such "vacant" periods represented a bolt for freedom by many state subjects and an improvement in human welfare.

What I wish to challenge here is a rarely examined prejudice that sees population aggregation at the apex of state centers as triumphs of civilization on the one hand, and decentralization into smaller political units on the other, as a breakdown or failure of political order. We should, I believe, aim to "normalize" collapse and see it rather as often inaugurating a periodic and possibly even salutary reformulation of political order. In the case of more centralized command-and-rationing economies such as Ur III, Crete, and Qin China, the problems were further compounded, and cycles of centralization, decentralization, and reaggregation seem to have been common.[31]

The "collapse" of an ancient state center is implicitly, but often falsely, associated with a number of human tragedies, such as high death toll. To be sure, an invasion, a war, or an epidemic may cause large-scale fatalities, but it is just as common for the abandonment of a state center to entail little if any loss of life. Such cases are better considered a redistribution of population, and, in the case of a war or epidemic, it is often the case that abandoning the city for the countryside spares many lives that would otherwise be lost. Much of the fascination with "collapse" comes to us from Edward Gibbon's *Decline and Fall of the Roman Empire*. But even in this classic case, it has been argued that there was no loss of population but rather a redistribution, as several non-Latin peoples, such as the Goths, were absorbed.[32] On a wider view, the "fall" of the Empire restored the "old regional patchwork" that had prevailed before the Empire was cobbled together from its constituent units.[33]

What is lost culturally when a large state center is abandoned or destroyed is thus an empirical question. Surely it is likely to have an effect on the division of labor, and scale of trade, and on monumental architecture. On the other hand, it is just as likely that the culture will survive—and be developed—in multiple smaller centers no longer in thrall to the center. One must never confound culture with state centers or the apex of a court culture with its broader foundations.

Above all, the well-being of a population must never be confounded with the power of a court or state center. It is not uncommon for the subjects of early states to leave both agriculture and urban centers to evade taxes, conscription, epidemics, and oppression. From one perspective they may be seen to have regressed to more rudimentary forms of subsistence, such as foraging or pastoralism. But from another, and I believe broader, perspective, they may well have avoided labor and grain taxes, escaped an epidemic, traded an oppressive serfdom for greater freedom and physical mobility, and perhaps avoided death in combat. The abandonment of the state may, in such cases, be experienced as an emancipation. This is emphatically not to deny that life outside the state may often be characterized by predation and violence of other kinds, but rather to assert that we have no warrant for assuming that the abandonment of an urban center is, ipso facto, a descent into brutality and violence.

The irregular cycles of aggregation and dispersal hark back to patterns of subsistence that predate the first appearance of states. Sharply colder and drier conditions in the Younger Dryas, for example, are reported to have driven previously dispersed populations toward warmer and wetter lowlands, where they aggregated to take advantage of a greater food supply. In Mesopotamia around 7,000 BCE (at the end of the Prepottery Neolithic Phase A), declining yields and perhaps disease seem to have prompted, by contrast, a general dispersal of population. Given high season-to-season variability in the timing and volume of rainfall, there is every reason to believe that agrarian peoples would have developed a repertoire in times of persistent hunger that called for dispersal from large settlements until conditions improved.[34] One scholar of Mesopotamian studies has suggested that the notion of an amphibious peasantry be extended across the usually sacred and impermeable boundary between farmers and pastoralists. As with Owen Lattimore's similarly radical suggestion for the Han-Mongol frontier in China, Adams believes that "the connection between nomads and sedentaries was a two-way street, with individuals and groups moving back and forth along this continuum as a response to environmental and social pressure."[35] What would seem to many to be a retrogression and civilizational heresy may on closer examination be nothing more than a prudent and long-practiced adaptation to environmental variability.

The sorts of adjustment designed to cope with, say, drought would have characterized any settled agrarian community at this time. We might call them non–state related oscillations to distinguish them from state effects. In the era of the earliest states, I believe, abandonment of the center was most often a direct or indirect effect of state formation. Given the unprecedented concentration of crops, people, livestock, and urban economic activity fostered by states, a whole series of effects—soil exhaustion, siltation, floods, salinization, epidemics, fire, malaria, none of which existed at anything like such levels before the state and any one of which could gradually or suddenly empty a city and destroy a state—were more common.

Finally, and perhaps most important for our purposes, was the direct political cause of state extinction: politicide! Crushing taxes in grain and labor, civil wars and wars of succession within the capital, intercity wars, oppressive measures of corporal punishment and arbitrary abuse may be called state effects, and they can singly or in combination bring about a state's collapse. The leakage of population away from the grain core and a persistent pattern of "heading for the hills" and pastoralism at a time of trouble might have served, in a state with an overriding concern for manpower, as a homeostatic device. Presumably, informed that numbers of its subjects were absconding, the state might have taken positive measures to lessen their burdens and stem the leakage. The frequency of collapse, however, suggests that the signals either were not received or were ignored.

Episodes of collapse are frequently succeeded by what comes to be known as a "dark age." Just as the meaning of collapse merits close and critical inspection, so the term "dark age" needs to be queried: "dark" for whom and in what respects? Dark ages are just as ubiquitous as storied dynastic highpoints of consolidation. The term is often a form of propaganda by which a centralizing dynasty contrasts its achievement with what it casts as the disunity and decentralization that preceded it. At a minimum, it seems unwarranted for the mere depopulation of a state center and the absence of monumental building and court records to be called a dark age and understood as the equivalent of the civilizational lights being extinguished. To be sure, there are in fact periods when invasions, epidemics, droughts, and floods do kill thousands and scatter (or enslave) the survivors. In such cases the term "dark age" seems appropriate as a point of departure. The "darkness" of the age, in any event, is a matter of empirical inquiry, not a label that can be taken for granted. The problem for the historian or archaeologist who seeks to illuminate a dark age is that our knowledge is so limited—that, after all, is why it's called a "dark age." At least two obstacles obscure our view. The first is that the self-reporting, and self-inflating, apex of an urban political formation has been removed. If we want to know what's going on, we will have to scout on the periphery, in the smaller towns, villages, and pastoral camps. Second, the trove of written records and bas reliefs has dwindled if not disappeared, and we are left if not exactly "in the dark," at best in the realm of oral culture that is hard to trace and date. The self-documenting court center that offered convenient one-stop shopping for historians and archaeologists is replaced by a fragmented, dispersed, and largely undocumented "dark age."

After the "collapse" of Ur III near the end of the third millennium BCE, the consensus holds that the Sumerian alluvium entered a "dark age," the duration of which is disputed. Many settled communities were deserted. "As sedentary life came near to flickering away, the local annals and archives which might have recorded this process seem to have disappeared altogether."[36] Of the magnitude of depopulation there is little doubt: "According to one estimate, south Levantine population crashed to a tenth or twentieth of its previous level," wrote Broodbank. "Most large settlements emptied out to be replaced by a scatter of tiny, short-lived sites."[37]

The usual reason given for the collapse was an "invasion" of Amorites, a pastoralist people perhaps driven from their homeland by a drought. There seems, however, not to have been great bloodshed—in keeping with our understanding of the importance of manpower—and Amorite hegemony seems to have been a gradual process. What happened to the population is a mystery. Perhaps it dispersed far and wide, but there is no evidence that the people were slaughtered. Another possibility is that the drought and/or an epidemic took many lives and scattered the survivors. Amorite rule, it seems, was more benign than that of Ur III. The Amorite rulers seem to have abolished most taxes and forced labor—perhaps to stem the hemorrhage of population—and encouraged a society of large farmers, merchants, and free subjects. It was, in any event, hardly a story of barbarian plunder and atrocities.

Most of the history of Mesopotamia that we have inherited comes from the more amply documented three-century "high-state" period of Ur III, Akkad, and Babylon's brief hegemony. We are reminded by Seth Richardson, however, that this period was anomalous and that seven centuries of the nine from 2,500 to 1,600 BCE were periods of division and decentralization.[38] There is no indication that this period, though "dark" in the sense of lacking a luminous, self-chronicling state, was in any sense dark in terms of famine or violence.

The first "dark age" of Egypt, called the First Intermediate Period, was slightly more than a century long (2,160–2,030 BCE), between the Old Kingdom and the Middle Kingdom. There seems not to have been any crash in population or even a radical dispersal of settlement patterns. Rather, it seems to have been a hiatus in the continuity of central rule. The apparent result was a rise of local provincial rulers—*nomarchs*—who now paid only nominal allegiance to the central court. Taxes may well have been reduced, while provincial elites availed themselves of the right to imitate the rituals previously reserved exclusively for the central elite. It represented a small democratization of culture. In sum, the First Intermediate Period seems less a dark age than a brief episode of decentralization touched off, almost certainly, by a period of low water levels in the Nile that led to crop failures and the loosening of the central state's grip on its subjects. Inscriptions from the period dwell as much on a revolution in social relations—on plunder, the looting of grain stores, the ascendance of the poor and destitution of the rich—as on deprivation in general.[39]

The dark age of Greece lasted roughly from 1,100 to 700 BCE. Many of the palatial centers were abandoned and often physically destroyed and burned; trade was vastly diminished, and writing in the Linear B script disappeared. The causes suggested are multiple and unverified: a Dorian invasion, invasion by mysterious "sea peoples" of the Mediterranean, drought, and perhaps disease. In terms of the culture it is seen as a dark age before the subsequent glories of Greece's Classical Age. But the oral epics of the *Odyssey* and the *Iliad*, as we have noted, date from precisely this dark age of Greece and were only later transcribed in the form in which we have come to know them. One might well argue, in fact, that such oral epics that survive by repeated performance and memorization constitute a far more democratic form of culture than texts that depend less on performance than on a small class of literate elites who can read them. While Greece's dark age represented a long and thorough eclipse of the earlier city-states, we know next to nothing about life in the smaller, fragmented, autonomous centers that survived, nor the role they may have played in laying the foundation for the subsequent flourishing of Classical Greece.

There may well be, then, a great deal to be said on behalf of classical dark ages in terms of human well-being. Much of the dispersion that characterizes them is likely to be a flight from war, taxes, epidemics, crop failures, and conscription. As such, it may stanch the worst losses that arise from concentrated sedentism under state rule. The decentralization that arises may not only lessen the state-imposed burdens but may even usher in a modest degree of egalitarianism. Finally, providing that we not necessarily equate the creation of culture exclusively with apical state centers, decentralization and dispersal may prompt both a reformulation and a diversity of cultural production.

I wish also to at least gesture in the direction of another unrecognized, undocumented true dark age far from state centers. Most of the world's population in the epoch of the early states comprised nonstate hunters and gatherers. William McNeill conjectures that they would have been demographically devastated when they came into contact with the novel diseases generated by concentrations in the grain core—diseases that for urban populations were becoming more endemic and hence less lethal.[40] If so, much of this nonstate population may have perished well outside of any documentation and notice—and therefore outside of recorded history—as was the case for the epidemiological devastation of New World populations as they succumbed to diseases that raced inland often well ahead of any European eyes. If we add to the toll of such diseases the scooping up of nonstate populations as slaves, a practice that continued into the nineteenth century, we have a "dark age" of epic proportions among peoples "without histories" that went unnoticed by history itself.

Course materials

Prof writing: The handbook of technical writing

Statistical methods: Introductory Business statistics

Microeconomics: microeconomics by McConnell, Brue, Flynn

CHAPTER SEVEN

The Golden Age of the Barbarians

The history of the peasants is written by the townsmen
The history of the nomads is written by the settled
The history of the hunter-gatherers is written by the farmers
The history of the nonstate peoples is written by the court
    scribes
All may be found in the archives catalogued under "Barbarian
    Histories"

LOOKED at from outer space in 2,500 BCE, the very earliest states in Mesopotamia, Egypt, and the Indus Valley (for example, Harrapan) would have been scarcely visible. In, say, 1,500 BCE there would have been a few more centers (Maya and the Yellow River), but their overall geographical presence may actually have shrunk. Even at the height of the Roman and early Han "superstates," the area of their effective control would have been stunningly modest. With respect to population, the vast majority throughout this period (and arguably up until at least 1600 CE) were still nonstate peoples: hunters and gatherers, marine collectors, horticulturalists, swidden-ers, pastoralists, and a good many farmers who were not effectively governed or taxed by any state.[1] The frontier, even in the Old World, was still sufficiently capacious to beckon those who wished to keep the state at arm's length.[2]

States, being largely agrarian phenomena, would, with the exception of some intermontane valleys, have looked like small alluvial archipelagoes, located on the floodplains of a handful of major rivers. Powerful as they might become, their sway was ecologically confined to the well-watered, rich soils that could support the concentration of labor and grain that was the basis of their power. Outside this ecological "sweet spot," in arid lands, in swamps and marshes, in the mountains, they could not rule. They might mount punitive expeditions and win an engagement or two, but rule was another thing. Most early states of any duration probably consisted of a directly ruled core region, a penumbra of peoples whose incorporation depended on the varying power and wealth of the state, and a zone quite outside its reach. For the most part, states did not seek to rule fiscally sterile areas beyond the core that would not normally repay the cost of governing them. Instead, states sought military allies and proxies in the hinterland and traded to obtain the scarce raw materials they needed.

The hinterland was not simply an *un*governed — or better put, a not-yet-governed — zone, but rather a zone governed, from the perspective of the state center, by "barbarians" and "savages." Though hardly precise Linnaean categories, "barbarians" often denoted a hostile pastoral people who posed a military threat to the states but who might, under certain circumstances, be incorporated; "savages," on the other hand, were seen as foraging and hunting bands not suitable as raw material for civilization, who might be ignored, killed, or enslaved. When Aristotle wrote of slaves as tools, one imagines that he had in mind "savages" and not all barbarians (for example, Persians).

The lens of "domestication" in general is useful for making sense of "barbarians" from the perspective of state centers. The grain growers and bondspeople at the state core are domesticated subjects, while foragers, hunters, and nomads are wild, savage, undomesticated peoples: barbarians. Barbarians are to domesticated subjects as wildlife, vermin, and varmints are to domesticated livestock. They are uncaptured at the very least and, at worst, represent a nuisance and threat that must be exterminated. In turn, weeds in the cultivated field are to domesticated crops as barbarians are to civilized life. They are a nuisance, and they and the birds, mice, and rats who appear uninvited at the harvest supper in the fields are a danger to the state and civilization. Weeds, varmints, vermin, and barbarians—the "undomesticated"—threaten civilization in the grain state. They must either be mastered and domesticated or, failing that, exterminated or rigorously excluded from the domus.

I should make it crystal clear, once again, that I am using the term "barbarian" in an ironic, tongue-in-cheek sense. "Barbarian" and its many cousins—"savage," "wild," "raw," "forest people," "hill people"—are terms invented in state centers to describe and stigmatize those who had not yet become state subjects. In the Ming Dynasty the term "cooked," referring to assimilating barbarians, meant, in practice, those who had settled, had been registered on the tax rolls, and who were in principle governed by Han magistrates—in short, those who were said to have "entered the map." A group that was identical in language and culture would often be divided into "raw" and "cooked" fractions entirely on the basis of whether they were outside or inside state administration. For the Chinese as for the Roman, the barbarians and tribes began precisely where taxes and sovereignty stopped. Let's understand, then, that henceforth, when I use the term "barbarian," it is merely an ironic shorthand for "nonstate peoples."

CIVILIZATIONS AND THEIR
BARBARIAN PENUMBRA

We have seen in great detail how the early state was radically unstable for internal structural, epidemiological, and political reasons. It was also vulnerable to predation from other states. But I wish to argue here that the threat posed by barbarians was perhaps the single most important factor limiting the growth of states for a period measured more in millennia than in centuries. From the Amorite incursions into Mesopotamia, through the Greek "dark age," the fragmentation of the Roman Empire, and the Yuan (Mongol) Dynasty in China, and perhaps beyond, the barbarian presence was the greatest danger to the state's existence and, at the very least, the crucial constraint on its growth.[3] I am speaking less of the barbarian "stars"—the Mongols, the Manchu, the Huns, the Mughals, Osman—than of the countless bands of nonstate peoples who gnawed relentlessly with raids on sedentary, grain-farming communities. Many of the nonstate, raiding peoples were themselves at least semisedentary: for example, Pathans, Kurds, Berbers.

The way we can best conceptualize this activity, I believe, is to see it as an advanced and lucrative form of hunting and foraging. Sedentary communities represented, for mobile foragers, an irresistible site for concentrated gathering. Some idea of the pickings they offered can be gained by this inventory of the loot from a large (ultimately unsuccessful!) hill raid on a lowland settlement in western India in late colonial times: 72 bullocks, 106 cows, 55 calves, 11 female buffaloes, 54 brass and copper pots, 50 pieces of clothing, 9 blankets, 19 iron ploughs, 65 axes, ornaments, and grain.[4]

The period between the first appearance of states and their hegemony over nonstate peoples represented, I believe, something of a "golden age of barbarians." What I mean is that it was in many ways "better" to be a barbarian *because* there were states—so long as those states were not too strong. States were juicy sites for plunder and tribute. Just as the state required a sedentary grain-growing population for its predations, so did this concentration of settled people, with their grain, livestock, manpower, and goods, serve as a site of extraction for more mobile predators. When the predator's mobility was enhanced by camels, horses, stirrups, or swift boats of shallow draft, the range and effectiveness of their raids was greatly extended. The returns to barbarian life would have been far less attractive in the absence of these concentrated foraging sites. If we think of the carrying capacity of barbarian ecology, my argument is that it was enhanced by the existence of petty states in much the same way that it would have been enhanced by a propitious stand of wild cereals or a migration of game. It would be hard to tell whether the microparasites of sedentary communities or the outbreaks of macroparasitic raiders contributed more to the limits on the growth of states and their populations.

Setting precise dates to the "golden age of barbarians" is surely a fool's errand. The history and geography of any particular area is likely to yield a very different configuration of state-barbarian relations, and one that is likely to shift over time. The Amorite "incursions" into Mesopotamia around 2,100 BCE may have represented a notable peak of barbarian "troubles," but it was surely not the only occasion on which the Mesopotamian city-states faced trouble from their hinterlands. And here we should recall that virtually all of our knowledge of barbarian "threats" comes from state sources— sources that might well have self-interested reasons to downplay or, more likely, to overdramatize the threat and to define the term "barbarian" narrowly or widely.

Conscious of the complexities, Barry Cunliffe bravely ventures to propose that, in the Mediterranean at least, the barbarian disruption of the ancient state world lasted for more than a millennium until 200 BCE. Within this period he identifies particularly the century between 1,250 and 1,150 BCE as the time when "the whole edifice of centralized, bureaucratic, palace-based exchange fell apart."[5] The virtual abandonment of many state centers at this time is often attributed to the so-called sea people invaders, perhaps of Mycenaean and Philistine origin, about whom little is known.[6] They raided Egypt in 1,224 BCE and again in 1,186 BCE, along with nomads from the desert to the west of the Nile. At about the same time, fortifications and towers proliferate in the northern Mediterranean, presumably to defend against raiders moving by land and by sea. Over the course of this long millennium a large proportion of the Mediterranean population had been displaced not once but several times. By the second century BCE, Cunliffe judges, "an all-pervading ethos of raiding had largely subsided," but not before the Celts had raided as far as Delphi.[7]

At the end of this period, on the other side of the Eurasian continent, the Qin and Han Dynasties were having their own troubles with the Xiongnu tribal confederacy over control of the lands within the large "Ordos loop" of the Yellow River. In the middle of the continent, Bennett Bronson claims that the relative absence of any strong states in the Indian subcontinent was due largely to the many powerful nomadic raiding groups that prevented states from consolidating. From the fourth century BCE until 1600 CE, "the entire northern two-thirds of the subcontinent produced exactly two moderately durable, region-spanning states: the [Chandra] Gupta and the Mughal," Bronson writes. "Neither of these nor any of the smaller northern states lasted longer than two centuries and anarchical interregna were everywhere prolonged and severe."[8]

Owen Lattimore, the pioneer of border studies in the context of China's relationship with its powerful, militarized, nomadic fringe to the north, sees a more general, continental pattern. He points to state walls and fortification against nonstate peoples springing up from western Europe through central Asia into China, and lasting until the Mongol invasions of Europe in the thirteenth century. It seems a rather extravagant claim, but, coming as it does from Lattimore, it merits pondering. "There was a linked chain of fortified northern frontiers of the ancient civilized world from the Pacific to the Atlantic. The earliest frontier walls appear to have been in the Iranian sector. The walled frontiers of the western Roman Empire in Britain and on the Rhine and Danube faced forest, upland, and meadow tribes, now pastoral nomads."[9]

The greatest boon that the appearance of states provided to barbarians, however, was less as sites for predation than as trading posts. Because states represented such narrow agro-ecologies, they relied on a host of products from outside the alluvium to survive. State and nonstate peoples were natural trading partners. As a state grew in population and wealth, so too did its commercial exchange with nearby barbarians. In the first millennium BCE there was a veritable explosion in seaborne commerce in the Mediterranean that exponentially increased the volume and value of trade. The greater part of the "barbarian economy" in this context was devoted to supplying lowland markets with raw materials and goods they required, much of which was in turn destined for reexport to other ports. A good part of what barbarians supplied was livestock in the most expansive sense of the term: cattle, sheep, and above all slaves. In return they received textiles, grain, iron- and copperware, pottery, and artisan luxury items, much of it too from "international" trade. Barbarian groups that controlled one or more of the major trading routes (usually a navigable river) to a major lowland center could reap large rewards and became, in turn, conspicuous sites of luxury, talent, and, if you will, "civilization."

Plunder of and trade with the state, then, made economic life on the state's margins more viable and lucrative than it could otherwise have been. But plunder and trade were not simply alternative modes of appropriation; as we shall see, they were very effectively combined in ways that mimicked certain forms of statecraft.

BARBARIAN GEOGRAPHY, BARBARIAN ECOLOGY

"Barbarians" are certainly not a culture or a lack thereof. Neither are they a "stage" of historical or evolutionary progress in which the highest stage is life in the state as taxpayer, in line with the historical discourse of incorporation shared by the Romans and Chinese. For Caesar incorporation meant moving from tribal (friendly or hostile) to "provincial" and perhaps eventually to Roman. For the Han it meant progressing from "raw" (hostile) to "cooked" (friendly) and perhaps eventually to Han. The intermediate steps "provincial" and "cooked" were specific categories of administrative and political incorporation to be followed, in ideal circumstances, by cultural assimilation. Put clinically and structurally, "barbarian" is best understood as a position vis-à-vis a state or empire. Barbarians are a people adjacent to a state but not in it. As Bronson puts it, they are simply "on the outside looking in."[10] Barbarians did not pay taxes; if they had a fiscal relationship with the state at all, they were expected to offer tribute as a collectivity.

Describing state geography and ecology in the ancient world is relatively easy on account of the agrarian and demographic requirements of state making. States were likely to arise only in rich, well-watered, bottomland soils. Until the last half of the first millennium BCE, when larger, sail-driven ships could transport larger cargoes longer distances, states had to hug the grain core quite tightly. Barbarian geography and ecology is, on the other hand, much harder to describe concisely because it constitutes a large and residual category; basically they comprise all those geographies that are unsuitable for state making. The barbarian zones most often referred to are the mountains and steppes. In fact, almost any area that was difficult to access, illegible and trackless, and unsuitable for intensive farming might qualify as a barbarian zone. Thus uncleared dense forest, swamps, marshes, river deltas, fens, moors, deserts, heath, arid wastes, and even the sea itself have been cast into this category by state discourse. A great many apparently ethnic names turn out to be, when translated literally, a description of a people's geography, applied to them by state discourse: "hill people," "swamp dwellers," "forest people," "people of the steppes." The only reason pastoral nomads of the steppe, mountain people, and sea people figure so prominently in state discourse about barbarians is that such peoples were not only out of reach but were also the most likely to pose a military threat to the state itself.

The figurative and often literal limit of a state's reach was often demarcated by a state-erected physical boundary between "civilized" and "barbarian" zones. The first great wall of this kind was the 250-kilometer-long "wall of the land" built around 2,000 BCE between the Tigris and Euphrates by command of Sumerian king Sulgi. Though it is typically described as a wall to keep the barbarian Amorites out (a task at which it failed), Anne Porter and others believe it had the additional purpose of keeping the southern Mesopotamian taxpaying cultivators in.[11] For the early Roman Empire, the barbarians "began" on the east bank of the Rhine, beyond which the Roman legions never ventured after their catastrophic defeat in the battle of Teutoburg Forest (9 CE). The Balkans, "a land of mountains and valleys cut by countless streams and with few large areas of flat land," were similarly marked by a boundary (*limes*) of fortifications.[12]

Barbarian geography corresponded with what is distinctive about barbarian ecology and demography. As a residual category it describes modes of subsistence and settlement that are not those of the state grain core. In a Sumerian myth, the goddess Adnigkidu is admonished not to wed a nomad god, Martu, as follows: "He who dwells in the mountains . . . having carried on much strife . . . he knows not submission, he eats uncooked food, he has no house where he lives, he is not interred when he dies . . ." One can scarcely imagine a more telling mirror image of life as a grain-producing, domus-based state subject.[13] The Record of Rites (Liji) of the Zhou Dynasty contrasts the barbarian tribes who ate meat (raw or cooked) instead of the "grain food" of the civilized. Among the Romans, the contrast between their diet of grain and the Gallic diet of meat and dairy products was a key marker of their claim to civilized status. Barbarians were dispersed and highly mobile, and lived in small settlements. They might be shifting cultivators, pastoralists, fisher folk, hunter-gatherers, foragers, or small-scale collector-traders. They might even plant some grain and eat it, but grain was unlikely to be their dominant staple as it was for state subjects. They were, by virtue of their mobility, their diverse livelihoods, and their dispersal, unsuitable raw material for appropriation and state building, and it was for precisely these reasons that they were called barbarians. Such distinctions admitted of differences in degree, and this, in turn, served to demarcate, for the state, those barbarians who were plausible candidates for civilization from those who were beyond the pale. To Roman eyes, the Celts, who cleared land, raised some grain, and built trading towns (*oppida*), were "high-end" barbarians, while acephalous, mobile hunting bands were irredeemable. Barbarian societies can, like the *oppida* Celts, be quite hierarchical, but their hierarchy is generally not based on inherited property and is typically flatter than the hierarchy found in agrarian kingdoms.

The vagaries of geography often meant that the central grain-core territory was fragmented by, say, hills and swamps, in which case the state's core might include several "unincorporated" barbarian areas. A state often bypassed or hopped over recalcitrant zones in the process of knitting together nearby arable areas. The Chinese, for example, distinguished between "inner barbarians," who were in such quarantined areas, and "outer barbarians," at the frontiers of the state. The civilizational narratives of the early states imply, if they don't state directly, that some primitives, through luck or cleverness, domesticated crops and animals, founded sedentary communities, and went on to found towns and states. They left primitivism behind for state and civilization. The barbarians, according to this account, are the ones who did not make the transition, those who remained outside. After this great divergence there were two spheres: the civilized sphere of settlement, towns, and states on the one hand and the primitive sphere of mobile, dispersed hunters, foragers, and pastoralists on the other. The membrane between the two spheres was permeable, but only in one direction. Primitives could enter the sphere of civilization—this was, after all, the grand narrative—but it was inconceivable that the "civilized" could ever revert to primitivism.

We now know this view to be, on the historical evidence, fundamentally wrong. It is mistaken for at least three reasons. First, it ignores the millennia of flux and movement back and forth between sedentary and nonsedentary modes of subsistence and the many mixed options in between. Fixed settlement and plough agriculture were necessary to state making, but they were just part of a large array of livelihood options to be taken up or abandoned as conditions changed. Second, the very act of establishing a state and its subsequent enlargement was itself typically an act of displacement. Some of the preexisting population may have been absorbed, but others, perhaps a majority, may have moved out of range. Many of a state's adjacent barbarian populations may well have been, in effect, refugees from the state-making process itself. Third, once states were created, as we have seen, there were frequently as many reasons for fleeing them as for entering them. If, as the standard narrative suggests, people are attracted to the state for the opportunities and security that it offers, it is also true that high rates of mortality coupled with flight from the state sphere were sufficiently offsetting that slaving, wars for capture, and forced resettlement seemed integral to the manpower needs of the early state.

The key point for our purposes is that, once established, the state was disgorging subjects as well as incorporating them. Causes for flight varied enormously—epidemics, crop failures, floods, salinization, taxes, war, and conscription—provoking both a steady leakage and occasionally a mass exodus. Some of the runaways went to neighboring states, but a good many of them—perhaps especially captives and slaves—left for the periphery and other modes of subsistence. They became, in effect, barbarians by design. Over time an increasingly large proportion of nonstate peoples were not "pristine primitives" who stubbornly refused the domus, but ex-state subjects who had chosen, albeit often in desperate circumstances, to keep the state at arm's length. This process, detailed by many anthropologists, among whom Pierre Clastres is perhaps the most famous, has been called "secondary primitivism."[14] The longer states existed, the more refugees they disgorged to the periphery. Places of refuge where they accumulated over time became "shatter zones," as their linguistic and cultural complexity reflected that they were peopled by various pulses of refugees over an extended period.

The process of secondary primitivism, or what might be called "going over to the barbarians," is far more common than any of the standard civilizational narratives allow for. It is particularly pronounced at times of state breakdown or interregna marked by war, epidemics, and environmental deterioration. In such circumstances, far from being seen as regrettable backsliding and privation, it may well have been experienced as a marked improvement in safety, nutrition, and social order. Becoming a barbarian was often a bid to improve one's lot.

Nomads, Christopher Beckwith has noted,

> were in general much better fed and led easier, longer lives than the inhabitants of the large agricultural states. There was

a constant drain of peoples escaping from China to the realms of the eastern steppe, where they did not hesitate to proclaim the superiority of the nomad lifestyle. Similarly, many Greeks and Romans joined the Huns and other Central Eurasian peoples, where they lived better and were treated better than they had been back home.[15]

Such voluntary self-nomadization was neither rare nor isolated. For China's Mongol frontier, Owen Lattimore, as noted earlier, has made the case most forcefully that the purpose of the Great Wall(s) was as much to keep the Chinese taxpayers inside as to block barbarian incursions and that, nonetheless, a great many taxpaying Han cultivators had "distanced themselves" from state space — especially during times of political and economic disorder — and "attached themselves quite readily to barbarian rulers."[16] Lattimore, as a student of frontiers in general, quotes a scholar of the late Western Roman Empire who noted the same pattern there too, as "the pitiless collection of taxes and the helplessness of citizens before wealthy law-breakers" drove Roman citizens to seek the protection of Attila's Huns.[17] "In other words," Lattimore adds, "there were times when the law and order of the barbarians was superior to those of civilization."[18]

Precisely because this practice of going over to the barbarians flies directly in the face of civilization's "just so" story, it is not a story one will find in the court chronicles and official histories. It is subversive in the most profound sense. The attraction of the Goths in the sixth century CE was at least as great as that of the Huns had been earlier. Totila (king of the Ostrogoths, 541–552 CE) not only accepted slaves and coloni into the Gothic army, but even turned them against their senatorial masters by promising them freedom and ownership of land. "In so doing he permitted and provided an excuse for something the Roman lower classes had been willing to do since the 3rd century": "to become Goths out of despair over their economic situation."[19]

A great many barbarians, then, were not primitives who had stayed or been left behind but rather political and economic refugees who had fled to the periphery to escape state-induced poverty, taxes, bondage, and war. As states proliferated and grew over time, they ground out ever greater numbers who voted with their feet. The existence of a large frontier—rather like migration to the New World for poor Europeans in the nineteenth and early twentieth centuries— provided a less dangerous avenue of relief than rebellion.[20] Without romanticizing life on the barbarian fringe, Beckwith, Lattimore, and others make it clear that leaving state space for the periphery was experienced less as a consignment to outer darkness than as an easing of conditions, if not an emancipation. As the state was weakened and under threat, the temptation was to press harder on the core to make good the losses which then risked further defections in a vicious cycle. A scenario of this kind, it appears, was partly to blame for the collapse of the Cretan and Mycenaean centralized palatial state (circa 1,100 BCE). "Under bureaucratic pressure to increase yield, the peasantry would despair and move away to fend for themselves, leaving the palace-dominated territory depopulated, much as the archaeological evidence suggests," Cunliffe writes. "Collapse would follow quickly."[21]

We return briefly to the imperative of manpower. The early state was successful to the extent that it could amass an appropriation zone consisting of grain growers packed together on productive soil. Holding that population in place or, failing that, replenishing losses was the key to statecraft. Confinement could help. "The only way to avoid losing population, power and wealth to central Eurasia was to build walls, limit trading at the frontier cities, and attack steppe peoples as often as necessary to destroy them or keep them away."[22]

Tribes are, in the first instance, an administrative fiction of the state; tribes begin where states end. The antonym for "tribe" is "peasant": that is, a state subject. That tribality is above all a relationship to the state is captured nicely by the Roman practice of reverting to the use of former tribal names to describe provincial populations that had broken away and rebelled against Rome. The fact that barbarians who menaced states and empires and therefore made it into the history books bear distinct names—Amorites, Scythians, Xiongnu, Mongols, Alamanni, Huns, Goths, Junghars—conveys an impression of cohesion and cultural identity that is usually wildly at odds with the facts. These groups were all loose confederacies of disparate peoples brought together briefly for military purposes and then characterized by the threatened state as a "people." Pastoralists in particular have remarkably flexible kinship structures, allowing them to incorporate and shed group members depending on such things as available pasture, number of livestock, and the tasks at hand—including military tasks. Like states, they too are typically manpower hungry and therefore quickly work refugees or captives into the lineage kinship structure.

For the Romans and the Tang Dynasty, tribes were territorial units of administration, having little or nothing to do with the characteristics of the people so designated. A great many of the so-called tribal names were simply place names: a particular valley, a range of hills, a stretch of river, a forest. In some cases the term might designate the character of the presumed group—for example, a group the Romans called Cimbri, which means "robbers" or "brigands." The aim of both the Romans and Chinese was to find or, failing that, simply to designate a leader or chief who would subsequently be responsible for the good behavior of his people. Under the Chinese system (*tusi*) of "using barbarians to rule barbarians" a tributary chief was appointed, given titles and privileges, and held accountable by Han officials for "his people." Over time, of course, such an administrative fiction might take on an autonomous existence of its own. Once in place the fictions were institutionalized by courts, tribute payments, lower native officials, land records, and public works, structuring that part of native life that involved contact with the state. A "people" originally conjured out of whole cloth by administrative fiat might come to adopt that fiction as a conscious, even defiant, identity. In Caesar's evolutionary scheme, described earlier, tribes preceded states. Given what we now know, it would be more accurate to say that states preceded tribes and, in fact, largely invented them as an instrument of rule.

RAIDING

After a raid by people from beyond the alluvium, a well-to-do resident of Ur wrote the following lament:

> He who came from the highland has carried my possessions to
> the highlands. . . . The swamp has swallowed my possessions.

. . . Men ignorant of silver have filled their hands with my silver. Men ignorant of gems have fastened my gems around their necks.[23]

While the density of grain, population, and livestock in a concentrated space is the source of a state's power, it is also the source of its potentially fatal vulnerability to mobile raiders.[24] To be sure, the state is often no richer than its periphery, but as we have seen, the decisive difference is that the wealth of the state, or any sedentary community, is all conveniently stacked up in a confined space, while the wealth of the periphery is widely dispersed. Mobile raiders, especially if they are mounted, have the military initiative. They can arrive at a time and place of their choosing and in sufficient numbers to overwhelm the weakest point of a settled community or to intercept a trading caravan. If they are numerous enough, they can take a fortified community. Their advantage lies in lightning raids; they are unlikely, for example, to lay siege to a fortified city, as the longer they stay put the longer a state has to mobilize against them, thus nullifying their tactical advantage. Under premodern conditions and perhaps even until the era of cannons, mobile armies of pastoralists have generally been superior to the aristocratic and peasant armies of states.[25] Even in regions without pastoralists and horses, the general pattern seems to be that more mobile peoples—hunter-gatherers, swiddeners, and boat people—tend to dominate and extract tribute from sedentary horticulturalists and farmers.[26]

The well-known Berber saying "Raiding is our agriculture," cited in my introduction, is significant. It gestures, I think, in the direction of an important truth about the parasitic quality of raiding. The granaries of a sedentary community may represent two or more years of agrarian toil that raiders can appropriate in a flash. Penned or corralled livestock are, in the same sense, living granaries that can be confiscated. And since the booty of a raid also typically included slaves to ransom, keep, or sell, they too represented a concentrated store of value and productivity—reared at considerable expense—that could be taken away in a day. From an even broader perspective, however, one might say that one parasite was displacing another, inasmuch as the raiders were confiscating and dispersing the accumulated assets of what had been, until then, a concentrated site of appropriation reserved exclusively for the state.[27]

Barbarian raiders were, for their part, relatively safe from retaliation by the state. Being mobile and dispersed, they could usually simply melt away, often into the hills, swamps, and trackless grasslands, where state armies followed at their peril. State armies might be effective against fixed objectives and sedentary communities but were largely helpless campaigning against acephalous bands with no central authority with whom to negotiate or to defeat in battle.

Another way of expressing the relative immunity of, say, Mongol raiders from Chinese counterattack is to note the absence, as Lattimore does, of nerve centers in the grasslands.[28] If we are to believe the words that Herodotus puts in the mouth of a Scythian interlocutor, nomad raiders were quite conscious of the military advantages of having no fixed property. "For we Scythians have no towns or planted lands, that we might meet you the sooner in battle, [otherwise] fearing that the one [town] be taken or the other [crops] be wasted."[29]

In the Mediterranean in the late second millennium BCE, the danger to states came less from grasslands and deserts than from the sea. Like the steppe or desert, the navigable sea offers unique opportunities for seaborne raiders to surprise coastal communities and sack them or, in some cases, to take them over as rulers. Sea nomads preyed on the huge growth in Mediterranean trade by piracy as well, the equivalent of the pastoralists preying on overland caravans. The king of Ugarit, near present-day Latakia in Syria, describes an attack on his kingdom when his own chariots and ships were absent: "Behold the enemy's ships came here; my cities were burned and they did evil things in my country"; "The seven ships of the enemy that came here inflicted much damage upon us."[30] In addition to their well-known attacks on Egypt and the Levant, naval raiders were probably responsible for the destruction of palatial Crete and the imperial Hittite heartland.[31] They were the precursors to other famed seaborne raiders such as the Vikings and the "sea gypsies" (*orang laut*) of Southeast Asia. Contemporary piracy in the Arabian Sea suggests that even today, speed, mobility, and surprise can, for a time at least, tactically prevail over "quasi-sedentary" container ships.

Little is known about the "sea pirates." They may well have often operated out of Cyprus and have been responsible for several waves of attacks over more than a century. Like pastoralist raiders, they were an extremely heterogeneous lot in terms of their cultural and linguistic backgrounds. In state documents and chronicles they appear as a source of terror and dread. Modern research, however, has rehabilitated them as not just raiders but city builders in many of the realms they captured.

There is a deep and fundamental contradiction to raiding that, once grasped, suggests why it is a radically unstable mode of subsistence, one that is likely under most circumstances to evolve into something quite different. Carried to its logical conclusion, raiding is self-liquidating. If, say, raiders attack a sedentary community, carrying off its livestock, grain, people, and valuables, the settlement is destroyed. Knowing its fate, others will be reluctant to settle there. If raiders were to make a practice of such attacks, they would, if successful, have killed all the "game" in the vicinity or, better put, "killed the goose that lays the golden egg." Much the same is true for raiders or pirates who attack caravans or shipping lanes. If they take everything, either the trade is extinguished or, more likely, it finds another, safer route.

Knowing this, raiders are most likely to adjust their strategy to something that looks more like a "protection racket." In return for a portion of the trade goods, harvest, livestock, and other valuables, the raiders "protect" the traders and communities against other raiders and, of course, against themselves. The relationship is analogous to endemism in diseases in which the pathogen makes a steady living from the host rather than killing it off. As there are likely to be a plurality of raiding groups, each group is likely to have particular communities it "taxes" and guards. Raiding, often quite devastating, still occurs, but it is most likely to be an attack by raiders on a community protected by another raiding community. Such attacks represented a form of indirect warfare between rival raiding groups. Protection rackets that are routine and that persist are a longer-run strategy than one-time sacking and therefore depend on a reasonably stable political and military environment. In extracting a sustainable surplus from sedentary communities and fending off external attacks to protect its base, a stable protection racket like this is hard to distinguish from the archaic state itself.[32]

Ancient states as a whole, in addition to building walls and raising armies of their own, often resorted to paying off powerful barbarians *not* to raid. The payments might take many forms. They might, to save face, be described as "gifts" in exchange for formal submission and tribute. They might consist in awarding a raiding group a monopoly over the control of trade in a particular location or over a particular commodity. They might be disguised as payment to a militia that would ensure peace at the border. In return for the payment, the raiders would agree to plunder only enemies of their allied state, and the state, for its part, would often recognize the raider's independence in a particular territory. Over time, if the arrangement lasted, the raider's protected zone might come to resemble a provincial, quasi-autonomous government.[33]

Relations between the (Eastern) Han Dynasty around 200 CE and its nomadic raiding neighbors, the Xiongnu, is an illuminating example of political accommodation. The Xiongnu would make lightning raids and retreat back to the steppes before state forces could retaliate. Soon afterward, the Xiongnu would dispatch envoys to the court promising peace in return for favorable terms for border trade or direct subsidies. The arrangement would be sealed by a treaty in which the nomads appear as tributaries and make the appropriate performance of allegiance in return for large subsidies. The "reverse" tribute was enormous: one-third of the annual government payroll went to buying off the nomads. Seven centuries later, under the Tang, officials were delivering half a million bolts of silk to the Uighurs annually on similar terms. On paper it may have looked as if the nomads were tributary inferiors to the Tang emperor, but the actual flow of revenue and goods suggests the opposite in practice. The nomads were, in effect, collecting bribes from the Tang in exchange for not attacking.[34]

One imagines that such protection rackets were more common than the documents allow, inasmuch as they were likely to be secrets of state which, if fully revealed, would risk contradicting the public facade of an all-powerful state. Herodotus notes that the Persian kings paid annual tribute to the Cissians (residents of Susa in the foothills of the Zagros Mountains at the edge of the Mesopotamian alluvium) lest they raid the Persian heartland and endanger its overland caravan trade. The Romans, after several defeats in the fourth century BCE, paid the Celts one thousand pounds of gold to prevent raiding, a practice they would repeat with the Huns and Goths.

If we step back and widen the lens, barbarian-state relations can be seen as a contest between the two parties for the right to appropriate the surplus from the sedentary grain-and-manpower module. It is this module that both is the basis for state formation and is equally essential for barbarian accumulation. It is the prize. One-time plunder raiding is likely to kill the host altogether, while a stable protection racket mimics the process of state appropriation and is compatible with the long-run productivity of the grain core.

TRADE ROUTES AND TAXABLE
GRAIN CORES

The earliest substantial communities were already dependent on trade and exchange with other ecological zones. The consolidation of larger states only increased this dependence. Given the early constraints on transportation, the juxtaposition in Mesopotamia and the Fertile Crescent of high plateau, intermontane valleys, piedmont steppe, and alluvium, along with navigable water, made possible a "vertical economy" of beneficial exchange.[35] Ur and Uruk were possible only by virtue of products from higher altitudes: stone, ores, oils, timber, limestone, soapstone, silver, lead, copper, grindstones, gems, gold, and, not least, slaves and captives. Most of these products were floated down watercourses. The longer and more navigable the river, the larger the potential polity. Smaller Mediterranean polities were miniature replications of this pattern. They were typically located on the alluvium of a river near the coast and on adjacent uplands and could thereby command trade and exchange for the whole watershed. "This combination was favored over time, thanks to its unrivalled ability to harness and integrate the food-mobilizing and wealth acquiring openings of both land and sea."[36]

The barbarian "stars" best known to history were no different in kind from earlier and smaller nonstate peoples—

hunters and gatherers, swiddeners, coastal foragers, herds-men—who raided small states and traded with them. What was unique was the unprecedented magnification of scale: of the confederations of mounted warriors, of the wealth of the lowland states, and of the volume and reach of trade. The em-phasis on raiding in most histories is understandable in view of the terror it evoked among elites of the threatened states who, after all, provide us with the written sources. This per-spective overlooks the centrality of trade and the degree to which raiding was often a means rather than an end in itself. Christopher Beckwith's emphasis on trade routes is illumi-nating:

> Chinese, Greek and Arab historical sources agree that the steppe peoples were above all interested in trade. The careful manner in which Central Eurasians generally undertook their conquests is revealing. They attempted to avoid conflict and tried to get cities to submit peacefully. Only when they re-sisted, or rebelled, was retribution necessary. . . . The Central Eurasians' conquests were designed to acquire trade routes or trading cities. But the reason for the acquisition was to secure occupied territory that could be taxed in order to pay for the rulers' socio-political infrastructure. If all this sounds exactly like what sedentary peripheral states were doing, that is be-cause it was indeed the same thing.[37]

The early agrarian states and the barbarian polities had broadly similar aims; both sought to dominate the grain-and-manpower core with its surplus. The Mongols, among other raiding nomads, compared the agrarian population to *ra'aya*, "herds."[38] Both sought to dominate the trade that was within reach. Both were slaving and raiding states in which the major booty of war and the major commodity in trade were human beings. In this respect they were competing protection rackets.

The linkage between raiding and trading is reflected in the Celtic fringe of the Roman Empire, particularly in Gaul. In Republican Rome, the Celts, as noted, were often paid off in gold for not raiding. Over time the Celtic towns (*oppida*) became, in effect, multiethnic trading posts along river routes to the Empire, dominating trade in that sector. In return for grain, oil, wine, fine cloth, and prestige goods, they might send raw materials, woolens, leather, salt pork, trained dogs, and cheeses to the Romans.[39]

The potential rewards for dominating land- and water-borne trade expanded exponentially as the trade itself expanded in the same fashion. That expansion had in part to do with technical factors such as improvement in boatbuilding, sail rigging, and navigation out of sight of the coast. Above all, of course, it depended on the substantial growth of both population and polities around the Mediterranean, the Black Sea, and the major rivers leading to them. Dating the expansion of trade is relatively arbitrary, but Barry Cunliffe notes that by around 1,500 BCE, major centers of population in Egypt, Mesopotamia, and Anatolia were major consumers of products from distant markets, and Crete had become a major naval power in the Mediterranean on the basis of that trade.[40] Three hundred years later the notorious "sea people" appeared to dominate the urban coastal centers of Cyprus and to have eclipsed the older agrarian states in the control of trade. Originally, trade in such treasured commodities as gold, silver, copper, tin, precious stones, fine textiles, cedar wood, and ivory had been monopolized, as far as possible, by the elites of the agrarian states.

But by 1,500 BCE that monopoly had been broken, and, in any event, the volume and variety of goods had swollen beyond recognition.

Trade over long distances was hardly new. Even before the Neolithic, valued commodities, so long as they were small and light, were exchanged over great distances: obsidian, precious and semiprecious stones, gold, carnelian beads. What was new was not so much the range of the trade but the fact that it had come increasingly to include bulk commodities moved long distances across the entire Mediterranean. Egypt became the "breadbasket" of the eastern Mediterranean, shipping grain to Greece and later to Rome. What is crucial as well is that the market for goods that were raised, grown, collected, and foraged *outside* the agrarian core had an exponentially larger potential market. Goods from the mountains, high plateaus, marine fringes, and marshes that might previously have circulated locally were now traded "worldwide." Beeswax and bitumen, used to caulk ships, were in great demand. Aromatic woods such as camphorwood and sandalwood, as well as aromatic resins such as frankincense and myrrh, were much prized. It would be hard to overestimate the importance of this transformation. Suddenly the periphery and semiperiphery of the early states were the sites of valuable commodities for which there was now an appreciable market. Foraging, hunting, and marine collecting became lucrative commercial activities.

A few brief analogies can help clarify what this shift meant. In the ninth century CE, with the growth of trade links between China and Southeast Asia, hunting and foraging in the forests of Borneo exploded. Some claim that the island, hitherto virtually unpopulated, was peopled by forest collectors hoping to take advantage of the trading opportunities in camphorwood, gold, hornbill ivory, rhinoceros horn, beeswax, rare spices, feathers, edible birds' nests, tortoise shells, and so on. A second analogy, much later, might be the worldwide demand for ivory—in the North Atlantic mainly for piano keys and billiard balls—that set off a myriad of intertribal wars for control of the trade and, not incidentally, destroyed much of the elephant population. The trade in beaver pelts in North America is another case. Today, the demand in the Chinese and Japanese market for ginseng root, caterpillar fungus, and matsutake mushrooms has made foraging a commercial activity that occasionally resembles the Klondike gold rush.[41] On a smaller but no less revolutionary scale for their epoch, the various peripheries of the agrarian states became valuable commercial landscapes—in some ways more valuable than the alluvium itself—thoroughly enmeshed in Mediterranean-wide trade networks. The possibilities for hunters, foragers, and marine collectors had never been more promising.

Central Eurasia had a wealth of products to trade for goods from the agrarian states, especially once shipping opened distant markets. Beckwith provides an extensive list of such products recorded by early travelers. The list is enormous, but an abbreviated version will illustrate its variety: copper, iron, horses, mules, furs, hides, wax, amber, swords, armor, fabrics, cotton, wool, carpets, blanket cloth, felt, tents, stirrups, bows, fine woods, linseed, nuts, and, never absent from the list, slaves.[42] Raiding by nomadic groups, which resembled warfare by agrarian states, is best understood as a means of acquiring tributary communities and of dominating the trade that circulated through them. It was not a result of nomadic poverty, still less a desire for shiny objects. All nomadic societies were complex in the sense that they practiced some agriculture as well as herding and had a substantial artisan class, so that they were not normally in need of staple cereals or technical expertise from the agrarian states.

The barbarians, broadly understood, were perhaps uniquely positioned to take advantage—and in many cases direct charge—of the explosion in trade. They were, after all, by virtue of their mobility and dispersion across several ecological zones, the connective tissue between the various sedentary cereal-intensive states. As trade grew, mobile nonstate peoples were able to dominate the arteries and capillaries of that trade and exact tribute for doing so. Mobility was, if anything, even more critical with respect to seaborne trade across the Mediterranean. These nomads of the sea were, one archaeologist explains, in all probability seamen who originally hired out their services to the established agrarian kingdoms in "official trade." As the scale of trade and its opportunities grew, they became an increasingly independent force capable of imposing themselves as coastal polities, raiding, trading, and exacting tribute on the model of their landward counterparts.[43]

DARK TWINS

State and nonstate peoples, agriculturalists and foragers, "barbarians" and "civilized" are twins, both in reality and semiotically. Each member of the pair conjures up its partner. And despite abundant historical evidence to the contrary, the peoples who have historically identified themselves as belong-

ing to the ostensibly more "evolved" member of each pair—
state people, agriculturalists, the "civilized"—have taken their
identity as essential, permanent, and superior. The most ten-
dentious of these pairs, the civilized-barbarian pair, are born
together as twins. Lattimore has articulated this "dark twin"
thesis most clearly:

> Not only the frontier between civilization and barbarism, but
> barbarian societies themselves, were in large measure cre-
> ated by the growth and geographical spread of the great an-
> cient civilizations. It is proper to speak of the barbarians as
> "primitive" only in that remote time when no civilization yet
> existed and when the forbearers of the civilized peoples were
> also primitive. From the moment civilization began to evolve
> . . . it recruited into civilization some of the people who had
> land and displaced others and the effect on those who were dis-
> placed [was] that . . . they modified their own economic prac-
> tices and experimented with new kinds of specialization and
> they also evolved new forms of social cohesion and political
> organization, and new ways of fighting. Civilization itself cre-
> ated its own barbarian plague.[44]

Although Lattimore ignores the millions of nonstate foragers, shifting cultivators, and marine collectors who were not
pastoralists, he does capture the parallel evolution of nomadism and states. These nomads, most especially those on horseback who "plagued" state centers, are best seen simply as the
strongest competitors of the state for control of the agrarian
surplus.[45] Hunters and gatherers or swiddeners might nibble
at the state, but politically mobilized large confederations of
mounted pastoralists were designed to extract wealth from
sedentary states; they were a "state in waiting" or, as Barfield
puts it, a "shadow empire."[46] In the most robust cases, such

as the itinerate state founded by Genghis Khan, the largest contiguous land empire in world history, and the "Comanche Empire" in the New World, we would be better advised to think of them as "horseback states."[47]

The relationship between a nomadic periphery and an adjacent state could take any number of forms and was, in any case, highly volatile. At the predatory end it might simply consist of occasional raids punctuated by retaliatory expeditions by state armies. Caesar's brutal campaigns in Gaul might be considered a rare example of a successful expedition that, despite many subsequent uprisings, extended Roman rule. In other cases, such as the Xiongnu, Uighurs, and Huns, the relationship might involve bribes, subsidies, and a kind of reverse tribute. Such arrangements, under which the barbarians received part of the proceeds of the sedentary grain complex in return for not raiding, might be thought of as a de facto joint sovereignty by state and barbarians. Under relatively stable conditions, such an equilibrium might approximate the frontier protection-racket model described earlier. Conditions, however, were rarely so stable with respect either to statecraft or to the often fragmented, fractious nomadic polity.

Two other "solutions" were possible, each of which, in effect, dissolved the dichotomy itself. The first was for the nomadic barbarians to conquer the state or empire and become a new ruling class. Such was the case at least twice in China's history—the Yuan and Manchu/Qing Dynasties— and with Osman, founder of the Ottoman Empire. The barbarians became the new elite of the sedentary state, living at the capital and operating the state apparatus. As the Chinese proverb has it, "You can conquer a kingdom on horseback, but to rule it, you have to dismount." The second alternative is far more common but less remarked upon, and that is for the nomads to become the cavalry/mercenaries of the state, patrolling the marches and keeping the other barbarians in check. In fact, it is the rare state or empire that has not recruited units from among the barbarians, often in return for trade privileges and local autonomy. Caesar's pacification of Gaul was accomplished largely with Gallic troops. In this case, rather than conquering the state, the barbarians became part of the military arm of an existing state along the lines of, say, the Cossacks or the Gurkhas. This pattern, in the colonial setting, has been called "indigenous sub-imperialism."[48] On a large scale the use of mercenaries poses its own risks for a sedentary state, as the Tang discovered when they, in effect, hired the Turkic Uighurs to suppress the huge An Lushan Rebellion.

The consensus among most "barbarian specialists" seems to be that nomadic pastoralists require sedentary communities as depots of manpower and revenue as well as trading outlets. Nomadic pastoralists have been known to forcibly resettle agricultural populations to create such depots. Furthermore, according to this view, barbarian confederations operate as "shadow empires" adjacent to and parasitic on large sedentary polities. Their quasi-derivative status is emphasized by the fact that they tend to disappear when their host collapses. As Nikolay Kradin puts it, "The degree of centralization among nomads is in direct proportion to the extent of the neighboring agricultural civilization. . . ."

The imperial and quasi-imperial organization of the nomads in Eurasia first developed after the ending of the "axial age" from the middle of the first millennium BCE at the time of the mighty agricultural empires (Qin in China, Maur in India, the Hellenistic states of Asia Minor, the Roman Empire in Europe) and in those regions . . . where the nomads were forced into contact with highly organized, agricultural, urban societies.[49]

Kradin and others include among the pairs that arise and fall together the Xiongnu and the Han, the Turkish Khaghanat and the Tang, the Huns and the Romans, the "sea people" and the Egyptians, and perhaps the Amorites and the Mesopotamian city-states. Presumably the Yuan and Manchu Dynasties do not count in this series, as they swallow the sedentary kingdom rather than disappearing.

It is all too characteristic, though no less deplorable, that so much ink is devoted to the barbarian states and the empires they bedeviled. Like a capital city that dominates the news, they dominate the historical coverage. A more even-handed history would chronicle the relationship of hundreds of smaller states with thousands of nearby nonstate peoples, not to mention the relation of predation and alliance between those nonstate peoples. In his account of Athens in the Peloponnesian Wars, for example, Thucydides discusses dozens of different hill and valley peoples: those with kings and without kings, those with whom Athens has relations of alliance, tribute, or enmity. Each of those pairs, were their histories known, would add immeasurably to our understanding of the relations between states and their nonstate neighbors.

A GOLDEN AGE?

There is, I believe, a long period, measured not in centuries but in millennia—between the earliest appearance of states and lasting until perhaps only four centuries ago—that might be called a "golden age for barbarians" and for nonstate peoples in general. For much of this long epoch, the political enclosure movement represented by the modern nation-state did not yet exist. Physical movement, flux, an open frontier, and mixed subsistence strategies were the hallmark of this entire period. Even the exceptional and often short-lived empires of this long epoch (the Roman, Han, Ming, and in the New World the Mayan peer polities and the Inka) could not impede large-scale population movements in and out of their political orbit. Hundreds and hundreds of petty states formed, thrived briefly, and decomposed into their elementary social units of villages, lineages, or bands. Populations were adept at modifying their subsistence strategies when circumstances dictated—abandoning the plough for the forest, the forest for swiddening, and swiddening for pastoralism. While the increase in population would have, by itself, encouraged more intensive subsistence strategies, the fragility of the state, its exposure to epidemics, and a large nonstate periphery would not have allowed us to discern anything like state hegemony until, say, 1600 CE at the earliest. Until then a large share of the world's population had never seen a (routine) tax collector or, if they had seen one, still had the option of making themselves fiscally invisible.

There is no particular need to insist on the quasi-arbitrary date of 1600 CE. It roughly marks the end of the great Eurasian barbarian waves: the seaborne Vikings from the eighth to the eleventh centuries, Tamerlane's great kingdom of the late fourteenth century, and the conquests of Osman and his immediate successors. Between them they destroyed, plundered, and conquered hundreds of polities large and small and displaced millions of people. They were also great slaving expeditions; among the major prizes of such campaigns were precious metals and human beings for sale. It is not so much that such raiding mixed with trade disappeared after 1600 CE as that it became more fragmented. Edward Gibbon, a comparatively rare voice with something to say on behalf of pagans, wondered whether there were any "barbarians" left in Europe in the late eighteenth century. (He might have considered the Barbary pirates, Macedonia, or the highland Scots, or have noticed that the Europeans had joined the Arabs in scouring the slaving ports of the African continent for slaves.) Outside Europe and the Mediterranean the pattern of raiding, trading, and slaving remained a major activity in the Malay world and in upland Southeast Asia among hill peoples. As states and durable gunpowder empires grew, the ability of nonstate peoples to raid and dominate small states shrank at a pace that depended greatly on the region and its geography.

The earliest states, because of the opportunities they opened for trade, supplemented by raiding and protection rackets, represented a qualitatively new environment for nonstate peoples. Now a good deal of the world around them was valuable; they could participate fully in the new opportunities for trade without becoming a subject of the state. There would have been periods when leaving behind the plough of a state subject to take up foraging, pastoralism, and marine collecting would have represented a rational economic calculation as well as a bolt for freedom. In such moments, it is likely that the proportion of barbarians vis-à-vis state subjects would have grown because life at the periphery had become more, not less, attractive.

The life of "late barbarians" would, on balance, have been rather good. Their subsistence was still spread across several food webs; being dispersed, they would have been less vulnerable to the failure of a single food source. They were more likely to be healthier and live longer—especially if they were female. More advantageous trade made for more leisure, thus further widening the leisure-drudgery ratio between foragers and farmers. Finally, and by no means trivial, barbarians were not subordinated or domesticated to the hierarchical social order of sedentary agriculture and the state. They were in almost every respect freer than the celebrated yeoman farmer. This is not a bad balance sheet for a class of barbarians over whom the waves of history were supposed to have rolled a long time ago.

There are, however, two deeply melancholy aspects of the golden age of barbarians. Each has directly to do with the ecologically given political fragmentation of barbarian life. Many of the trade goods brought to the trading states were, of course, other nonstate peoples who could be sold into bondage at the state core. So pervasive was this practice in mainland Southeast Asia that one can identify something like a chain of predation in which more strategically located and powerful groups raided their weaker and more dispersed neighbors. In so doing they reinforced the state core at the expense of their fellow barbarians.

The second melancholy aspect of the new livelihoods at the periphery afforded by states was, as previously noted, the sale of their martial skills to states as mercenaries. One would be hard put to find an early state that did not enlist nonstate peoples — sometimes wholesale — in their armies, to catch runaway slaves, and to repress revolts among their own restive populations. Barbarian levies had as much to do with building states as with plundering them. By systematically replenishing the state's manpower base by slaving and by protecting and expanding the state with its military services, the barbarians willingly dug their own grave.

Notes

INTRODUCTION

1. The term was first coined by the Dutch climate scientist Paul Crutzen in 2001.

2. For the dating, personal communication, David Wengrow.

3. It's hard to avoid asking oneself, "Where did we go wrong to end up here?" That question is far too ambitious for me to tackle. One thing stands out, however, and that is that our trouble is largely of our own making. This, in turn, suggests a medical analogy. More than two-thirds of hospitalizations in industrial countries, it is claimed, are for iatrogenic illnesses: medical conditions that result from previous medical interventions and therapy. One might say that our current environmental ills are largely iatrogenic. If so, the first step is perhaps to elicit a long and deep medical history that might help us trace the origins of our current complaints.

4. In the first millennium BCE—later than the period on which I focus—when nomadic pastoralism is combined with the rearing of horses, a new kind of nonsedentary, grassland empire becomes possible, exemplified by the Mongols and, much later in the New World, by the Comanche. For such unique polities see, Pekka Hämäläinen, "What's in a Concept? The Kinetic Empire of the Comanches," *History and Theory* 52, no. 1 (2013): 81–90, and Mitchell, *Horse Nations*.

5. The only sensitive exploration of this topic I know of is Bruce

Chatwin's fine book written about Australia, *The Songlines* (London: Cape, 1987). The Roma, aka Gypsies, are a modern example of determined mobility—so much so that the famous Norwegian diplomat Fridtjof Nansen proposed after World War II issuing them what would have been the first "European" passports.

6. Urban populations, before the revolution in sanitation (sewage and clean water) of the mid-nineteenth century and before vaccination and antibiotics, generally had such high rates of mortality that they grew only by large-scale in-migration from the countryside.

7. In fact, it seems that such sites of wild stands and/or cultivated but nondomesticated grains and the periodic gatherings to harvest the grains and store them were common enough for them to be *misinterpreted* as permanent, sedentary communities cultivating fully domesticated crops. See in this connection the careful argument of Asouti and Fuller, "Emergence of Agriculture in Southwest Asia."

8. For perhaps the best and most detailed summaries of the current state of knowledge, see Fuller et al., "Cultivation and Domestication Has Multiple Origins," and Asouti and Fuller, "Emergence of Agriculture in Southwest Asia."

9. Algaze, "Initial Social Complexity in Southwestern Asia."

10. A good many nomadic peoples did have scripts (often borrowed from sedentary peoples), but they typically wrote on perishable material (bark, bamboo leaves, reeds) and for nonstate purposes (such as memorizing spells and love poetry). The heavy clay tablets of the southern alluvium of Mesopotamia are decidedly the writing technology of a sedentary people, and that is why so much of it survives.

11. Carneiro, "A Theory of the Origin of the State."

12. See McAnany and Yoffee, *Questioning Collapse.*

13. See Thomas J. Barfield, *The Perilous Frontier: Nomadic Empires and China* (Oxford: Blackwell, 1992).

I. THE DOMESTICATION OF FIRE, PLANTS, ANIMALS, AND . . . US

1. C. K. Brain, *The Hunters or the Hunted? An Introduction to African Cave Taphonomy* (Chicago: University of Chicago Press, 1981), cited in Goudsblom, *Fire and Civilization.*

2. Cronon, *Changes in the Land.*

3. For this still disputed contention, see William Ruddiman, "The Anthropogenic Greenhouse Era Began Thousands of Years Ago," *Climatic Change* 16 (2003): 261–293, and R. J. Nevle et al., "Ecological-Hydrological Effects of Reduced Biomass Burning in the Neo-Tropics After AD 1600," *Geological Society of America Meeting*, Minneapolis, October 11, 2011, abstract.

4. Zeder, "The Broad Spectrum Revolution at 40." Although I concentrate here on fire as a tool for landscape modification, hunting, and cooking, fire was used as a tool for hardening wooden tools, for splitting stones, for shaping weapons, and for raiding beehives long before the Neolithic revolution. See Pyne, *World Fire*.

5. Jones, *Feast*, 107.

6. Wrangham, *Catching Fire*, 40–53.

7. At this point a reader might ask why it was that Homo sapiens was a more successful invasive than Homo neanderthalensis, who, after all, had fire and cooking as well. One answer, different from that of higher fertility, is proposed by Pat Shipman. She suggests that the decisive difference rests with another tool, the domesticated wolf that allowed Homo sapiens to become a vastly more efficient hunter of big game rather than largely a scavenger. She makes a persuasive case that "wolf-dogs" had been tamed—or had attached themselves to Homo sapiens—more than thirty-six thousand years ago, when the two hominids lived in close proximity. She claims that this was also the time when most large game animals, owing to Homo sapiens' use of dogs for hunting, were in steep decline or extinct. Much of her argument hinges on the disputed temporal and spatial overlap of the two subspecies and the hunting grounds they contested. Why Homo neanderthalensis did not then also domesticate the wolf is a mystery to me. See *The Invaders*.

8. For both fire and cooking, see Goudsblom, *Fire and Civilization*, and Wrangham, *Catching Fire*.

9. Anders E. Carlson, "What Caused the Younger Dryas Cold Event," *Geology* 38, no. 4 (2010): 383–384, http://geology.gsapubs.org /content/38/4/383.short?rss=1&ssource=mfr. Although the dating of the beginning of the Younger Dryas and Lake Agassiz's turn east from the Mississippi drainage do not quite match, it does seem likely that some pulse of glacial melt was responsible for the cold snap.

10. Zeder, "The Origins of Agriculture."

11. Pournelle, "Marshland of Cities." For subsequent, but more truncated, versions of her findings see Pournelle, Darweesh, and Hritz, "Resilient Landscapes"; Hritz and Pournelle, "Feeding History." Pournelle's thesis is foreshadowed—but with far less hard evidence—by others, for example, Pollock, *Ancient Mesopotamia*, 65–66; Matthews, *The Archaeology of Mesopotamia*, 86. For a deeper historical and geological view, as well as a recasting of Gordon Childe's "oasis theory of civilization," see Rose, "New Light on Human Prehistory."

12. See, among others, Pollock, *Ancient Mesopotamia*, 32–37.

13. The process is beautifully described by Azzam Awash as follows: "It was not coincidental that agriculture first developed in the natural renewable fertility of the grasslands surrounding the marshes. What the Sumerians did was invent an ingenious irrigation system which their Marsh Arab inheritors continued using. Following the peak of the floods, they broadcast seeds on the higher lands that first start emerging as the floodwaters recede. These higher lands get covered twice a day as a result of the tidal actions of the Gulf that slows the flow in the Tigris and Euphrates causing a 'backup' of the water. The seeds thus get irrigated automatically without having to open canals or pump water. As the seedlings grow, however, the water recedes too far to allow for irrigation, and thus the seedlings are transplanted from the higher land into the low lying fields/grasslands. The irrigation system continues to provide water twice a day well into the early days of summer. By the time the floodwaters have receded, the roots of the seedling would tap into the groundwater and are in no need of the hard labor of irrigation." "The Mesopotamian Marshlands: A Personal Recollection," in Crawford, *The Sumerian World*, 640.

14. Latin American specialists will recognize the similarities between this pattern of adjacent ecological zones and subsistence security with the concept of a "vertical archipelago" of ecological zones in the Andean state made famous by John V. Murra. See, for example, Rowe and Murra, "An Interview with John V. Murra."

15. Sherratt, "Reviving the Grand Narrative," 13.

16. Heather, *The Fall of the Roman Empire*, 111.

17. H. R. Hall, *A Season's Work at Ur, Al-Ubaid, Abu-Shahrain (Eridu) and Elsewhere* . . . , quoted in Pournelle, "Marshland of Cities," 129.

18. For a perceptive analysis of this process and this logic, see D'Souza, *Drowned and Dammed.*

19. Smith, "Low Level Food Production."

20. Zeder, "The Origins of Agriculture," S230–S231.

21. Zeder, "After the Revolution," 99.

22. Endicott, "Introduction: Southeast Asia," 275. Endicott and Geoffrey Benjamin term this shift "respecialization."

23. Febvre, *A Geographical Introduction to History*, 241.

24. The term is used by Ian Hodder in *The Domestication of Europe.* Although I find Hodder's concept of the "domus" helpful to think with, the late Andrew Sherratt was quite correct to observe that "a will to sedentism" could not be posited as a causal force in human affairs. See Sherratt, "Reviving the Grand Narrative," 9–10.

25. Porter, *Mobile Pastoralism*, 351–393.

26. The question of "storage," including "social storage" and reciprocity as a means to cope with a variable environment, is examined from many angles in Halstead and O'Shea, *Bad Year Economics.*

27. For a careful analysis, see Rowley-Conwy and Zvelibil, "Saving It for Later."

28. Park, "Early Trends Toward Class Stratification."

29. As with many ideas, I discovered that this one too was not original with me! See Manning, *Against the Grain*, 28.

2. LANDSCAPING THE WORLD

1. Zeder, "Introduction," 8. Zeder claims that there is evidence for humans "actively tilling and tending wild stands of einkorn and rye at both Abu Hureyra and nearby Mureybet during the late epi-Paleolithic 15,000–13,000 BCE." For a documented and enlightening view of the transition from hunting and gathering to fixed-field cultivation, see Moore, Hillman, and Legge, *Village on the Euphrates.*

2. Moore, Hillman, and Legge, *Village on the Euphrates*, 387. The authors point to the "now dominant weeds of dry cereal cultivation"—clovers, medicks, and wild fenugreek relatives, a wall barley, small-seeded grasses, twitches, and gromwell (bugloss family)—that appear in quantity in the Middle East in ancient seed remains, which they label a sure sign of cultivation.

3. Lest one think such heroics are confined to Homo sapiens, the

little fish-eating auk managed, by colonizing northern Greenland in large numbers, to create enough soil with its wastes to create an attractive habitat for small mammals whose presence, in turn, attracted larger predators, including the polar bear.

4. See Catherine Fowler, "Ecological/Cosmological Knowledge and Land Management Among Hunter-Gatherers," in Lee and Daly, *The Cambridge Encyclopedia of Hunters and Gatherers*, 419–425.

5. Boserup, *The Conditions of Agricultural Growth*.

6. For the most remarkable and brilliantly illustrated survey of the origins of agriculture with an emphasis on trade, see Sherratt, "The Origins of Farming in South-West Asia."

7. I ignore, in this context, the weedy escapees, rather like pigs, that do manage to thrive outside the domus: oats, rye, vetch, false flax, carrot, radish, and sunflower.

8. Diamond, *Guns, Germs, and Steel*, 172–174.

9. Of the first four-footed domesticates, the pig and the goat can and have slipped easily from the domestic sphere to "ferality" with remarkable success.

10. For an extended development of the domus in the context of Europe, see Hodder, *The Domestication of Europe*.

11. For the Berlaev experiments, see Trut, "Early Canine Domestication."

12. Zeder, "Pathways to Animal Domestication."

13. Zeder et al., "Documenting Domestication," and Zeder, "Pathways to Animal Domestication."

14. R. J. Berry, "The Genetical Implications of Domestication in Animals," in Ucko and Dimbleby, *The Domestication and Exploitation of Plants and Animals*, 207–217.

15. See T. I. Molleson, "The People of Abu Hureyra" in Moore, Hillman, and Legge, *Village on the Euphrates*, 301–324.

16. Leach, "Human Domestication Reconsidered."

17. The preeminent theorist of the domus as the key social unit of agrarian society is Ian Hodder. The central role he assigns the domus in the process of domestication in *The Domestication of Europe* is prefigured by Peter J. Wilson in *The Domestication of the Human Species*.

18. Leach, "Human Domestication Reconsidered," 359.

19. Two common candidates for adaptations are the appearance of

the sickle cell trait as protection against malaria, which had become epidemic owing to human changes in cultivated landscapes, and the rise of lactose tolerance, especially among pastoral nomads. More controversial are the interpretations of when blood types A, B, and AB developed and from what epidemic diseases they appear to offer some protection. See, in general, Boyden, *The Impact of Civilisation on the Biology of Man.*

20. Pollan, *The Botany of Desire*, xi–xiv.

21. Evans-Pritchard, *The Nuer,* 36.

22. See Conklin, *Hanunòo Agriculture,* and Lévi-Strauss, *La Pensée sauvage.*

23. Owen Lattimore, comparing the Mongol pastoralist with the Han farmer, puts the matter more strongly that I would, having, as a mediocre farmer, understood how complex it is to master. "As a matter of fact the Mongol, trained from childhood to be independent and to do all kinds of different things for himself, to work leather and felt, to drive a cart and handle a caravan, to be out in all weather and find his way over great distances and above all to make his own decisions for himself, promptly and in every kind of circumstance ought to be well-placed in competition with the peasant colonist who has lived in one mud hut all his life, attending without any exercise of initiative to an unchanging routine of planting and harvesting with his decisions made for him by his landlord and the calendar." "On the Wickedness of Being Nomads," quotation on 422.

24. Elias, *The Civilizing Process.*

25. Tocqueville, *Democracy in America,* 2: 1067.

3. ZOONOSES

1. Moore, Hillman, and Legge, *Village on the Euphrates,* 393. This is an amazingly comprehensive and valuable survey of the richest site in Mesopotamia.

2. Burke and Pomeranz, *The Environment and World History,* 91, citing Peter Christensen, *The Decline of Iranshahr.* The period Christensen is referring to falls later, but he dates the origin of such diseases to the Neolithic transition itself. See chapter 7 and pp. 75 ff.

3. It is quite possible that advances in the recovery of genetic material will soon provide more robust evidence for such suspicions.

4. See, among others, Porter, *Mobile Pastoralism,* 253–254; Rad-

ner, "Fressen und gefressen werden"; Karen Radner, "The Assyrian King and His Scholars: The Syrio-Anatolian and Egyptian Schools," in W. Lukic and R. Mattila, eds., *Of Gods, Trees, Kings, and Scholars: Neo Assyrian and Related Studies in Honour of Simo Parpola, Studia Orientalia 106* (Helsinki, 2009), 221–233; Walter Farber, "How to Marry a Disease: Epidemics, Contagion, and a Magic Ritual Against the 'Hand of the Ghost,'" in H. F. J. Horstmanshoff and M. Stol, eds., *Magic and Rationality in Ancient Near Eastern and Graeco-Roman Medicine* (Leiden: Brill, 2004), 117–132.

5. Farber, "Health Care and Epidemics in Antiquity." Evidence here comes largely from Mari on the Euphrates from Uruk around the early second millennium BCE.

6. Nemet-Rejat, *Daily Life in Ancient Mesopotamia*, 80.

7. Ibid., 146. Nemet-Rejat adds, "An omen reported plague gods marching with the troops, most likely a reference to typhus."

8. See especially Groube, "The Impact of Diseases"; Burnet and White, *The Natural History of Infectious Disease*, especially chapters 4–6; and McNeill, *Plagues and People.*

9. McNeill, *Plagues and People*, 51.

10. Polio is an example of an epidemic related to an excess of hygiene. In a major city in the global south like Bombay, for example, an overwhelming percentage of the children under five will have polio antibodies in their system, showing that they have been exposed to the disease, which is spread by feces and is rarely fatal to infants. For one not exposed at an early age, however, the disease contracted later in life is far more severe.

11. Moore, Hillman, and Legge, *Village on the Euphrates*, 369.

12. Roosevelt, "Population, Health, and the Evolution of Subsistence."

13. Nissen and Heine, *From Mesopotamia to Iraq.*

14. Dark and Gent, "Pests and Diseases of Prehistoric Crops."

15. Ibid., 60.

16. See Lee, "Population Growth and the Beginnings of Sedentary Life."

17. See Redman, *Human Impact on Ancient Environments*, 79 and 169, where he notes that a small change in the age of first conception or a reduction by three or four months in the interval between conceptions

can, over time, make a huge difference in population growth rates. A hypothetical band of one hundred growing at a rate of 1.4 percent—that is, doubling every 50 years—would, in a mere 850 years, number thirteen million.

18. In Europe itself, it seems that only 20–28 percent of the DNA of early farmers can be traced to migration from the Near East cradles of agriculture. This implies, then, that the great bulk of early farmers were the descendants of indigenous hunter-gatherers. See Morris, *Why the West Rules—for Now*, 112.

4. AGRO-ECOLOGY OF THE EARLY STATE

Epigraphs: Sumerian text quoted in Tate Paulette, "Grain, Storage, and State-Making," 85; Lawrence, Preface to Dostoevsky's "The Grand Inquisitor."

1. Pournelle, "Marshland of Cities," 255.

2. Pournelle, "Physical Geography," 28.

3. Pournelle and Algaze, "Travels in Edin," 7–9.

4. Sumerian irrigation, where it was practiced, is now judged to have been far less centralized than previously thought, with the shorter canal work being readily organized by local communities. See Wilkinson, "Hydraulic Landscapes and Irrigation Systems," 48. The same, it appears, was the case in Egypt as well.

5. The question of what precisely constitutes an army is not simple. In early Mesopotamia there are depictions of battles, weapons, armor, and, of course, booty and prisoners from campaigns. The texts make clear that there were both conscription and widespread efforts to avoid it. The first clear textual reference to a standing army, however, comes later under the Akkadian dynast Sargon (2,334–2,279 BCE); Nemet-Rejat, *Daily Life in Ancient Mesopotamia*, 231.

6. Nissen, *The Early History of the Ancient Near East*, 127. Definitive archaeological evidence for elite burials occurs later, around 2,700 BCE, and evidence for kings and standing armies only around 2,500 BCE. As there are few documented burials at all before 2,700 BCE, the adage "Absence of evidence is not evidence of absence" applies.

7. Nissen and Heine, *From Mesopotamia to Iraq*, 42.

8. Postgate, "A Sumerian City," 83.

9. Nissen, *The Early History of the Ancient Near East*, 130.

10. Nemet-Rejat, *Daily Life in Ancient Mesopotamia*, 100.

11. As trade developed later during the second millennium BCE, strategic chokepoints on overland and riverine trade routes—places without a rural hinterland—could serve as places of state making. Much later, with the sea transport of bulk commodities, state building at privileged nodes of trade (Venice, Genoa, Amsterdam) might give birth to maritime states receiving much of their food supply by waterborne transport from considerable distances.

12. Owen Lattimore, "The Frontier in History," 475.

13. The copper and tin would have been semiprocessed, as the alluvium lacked the high-quality fuel required to smelt.

14. The obvious exceptions would be the natural "chokepoints" on overland trade routes, such as mountain passes and fords and desert oases. The Straits of Melaka, an important node of state formation in Southeast Asia, is a classic example of both water transport routes and a chokepoint, in this case commanding the early India-China maritime trade route.

15. This assertion, which I distinctly recall reading in the opening paragraphs of a history of nineteenth-century Britain, was challenged by one of my readers as a possible "urban myth." Although I have not been able to retrieve the original citation, I can document the assertion in more substantial ways. The assertion is not quite true but nearly so! A relatively fast stagecoach (before Macadam!) was likely to average 20 miles a day. The distance from London to Edinburgh is about 400 miles; hence the trip would take about twenty days. Again, this is before new road-building and maintenance routines were perfected, when speeds were slow and night travel avoided. By 1830, however, with better roads, nonstop night and day travel and *fifty* changes of horses at lightning speed, the trip could be made in forty-eight hours. Most travelers would have preferred sea passage from London to Edinburgh. The distance from Southampton to Cape Town is roughly 7,000 nautical miles. While the fastest clipper ships, built of course for speed, might, in 1800, travel as much as 460 miles in a single day, a more realistic average speed for a clipper ship might be 300 miles per day. At that speed the trip would take twenty-three or twenty-four days. In more general terms costs by water in preindustrial Europe were estimated by one authority to be one twentieth of overland transportation costs. For example, an overland shipment of coal in the sixteenth century lost 10 percent of its value per mile, thus making coal

shipments longer than 10 miles profitless. Grain shipments, having more value per unit of weight and volume, lost only 0.4 percent of their value per mile traveled, permitting shipment of up to 250 miles before they became a losing proposition. Of course, questions of predation (highwaymen, brigands, pirates), and therefore of armed escorts, would reduce appreciably these abstract econometric calculations. See Meir Kohn, "The Cost of Transportation in Pre-industrial Europe," chapter 5 of *The Origins of Western Economic Success: Commerce, Finance, and Government in Pre-industrial Europe,* January 2001, http://www.dartmouth.edu/~mkohn /orgins.html, 50–51; http://ports.com/sea-route/port-of-southampton ,united-kingdom/port-of-cape-town,south-africa/.

16. Geographic barriers are important in still another respect. Inasmuch as the state requires an abundant population—as cultivators, laborers, soldiers, taxpayers—it helps if they have nowhere to run away to if they become dissatisfied. As Robert Carneiro argued for Mesopotamia, the population was hemmed in, or in his term circumscribed— one might as well say trapped—by a frontier of marshes, sea, arid lands, and mountains so that there was no easy way grain farmers could move away from the state. Would-be state makers had, he argued, a nearly captive population. He argued similarly for the Egyptian and early Yellow River states, bordered by deserts, as compared, say, with the Amazonian Basin or the eastern woodlands of North America. Although there is ample evidence historically of people moving from agriculture to pastoralism, to swiddening, to maritime livelihoods, and even to hunting and gathering, the existence of both geographic and ecological barriers and perhaps hostile peoples makes it easier for pristine states to hold their population on the alluvium. The problem for the Mesopotamian case is that it was relatively easy for agriculturalists to move into pastoralism when desirable and, for that matter, to move northward in the alluvium along the Tigris and/or Euphrates Valleys. Carneiro, "A Theory of the Origin of the State."

17. Once again, I am not referring here to the first sedentism but rather to the first durable populated settlements that later gave rise to the first states. The first sedentism in the alluvium was, here as elsewhere, a nonagricultural sedentism based on foraging and hunting at the seams of adjacent ecosystems with abundant resources. Perhaps the first sedentary communities in the world belonged to the coastal Jōmon

culture of northeast Japan which was, at 12,000 BCE, contemporaneous with and likely earlier than the Natufian period in the Fertile Crescent. Like the ecosystem described by Pournelle, the rich marine and woodland environment amid which the Jōmon foraged was, like that of the native Americans in the Pacific Northwest, close at hand.

18. Pournelle, "Marshland of Cities," 202.

19. The Andean crops amaranth and quinoa, in the same family of "pseudocereals," seem not to have figured as major tax crops, perhaps because their seeds ripen irregularly over a long period. Personal communication, Alder Keleman, September 2015.

20. Febvre, *A Geographical Introduction to History*, part III, 171-200.

21. See the parallel argument by Manning, *Against the Grain*, chapters 1 and 2.

22. As most of the plant nutrients for irrigated rice are delivered in the irrigation water rather than by the soil, such rice cultivation requires less fallowing or animal manure than, say, wheat or maize cultivation to be sustainable for long periods.

23. I elaborated this argument about the political implications of tuber and root cultivation on the one hand and cereal cultivation on the other at great length in *The Art of Not Being Governed*, 64-97, 178-219. Here I distinguished "state" crops like rice and "state-evading" crops like cassava and potatoes. I argued both that states depended on grain crops on fixed fields and that populations wishing to evade taxation and state control adopted subsistence strategies such as root crops, swidden—shifting—cultivation, hunting, and foraging to place themselves outside of state control. More recently a similar but not identical argument has been made by J. Mayshar et al., "Cereals, Appropriability, and Hierarchy." The authors note the key difference in appropriability between cereals and roots and tubers, although they fail to see that in many settings what is planted may be a political choice and that embryonic states encourage and often mandate cereal cultivation. While Mayshar et al. correctly associate cereal grains with state and hierarchy and root crops with non-state, egalitarian societies, they wrongly take subsistence strategies as a primordial given and not the product of political institutions and political choice. Wherever there is adequate water and decent soil, many choices are possible. The authors further assert—apparently on the basis solely of institutional economics' theory of the provision of public goods—that

state creation is a benign, elite-initiated invention to defend the community's stored grain against "robbers." My view, by contrast, is that the state originated as a protection racket in which one band of robbers prevailed. While I am delighted to know that others have detected the important relationship between cultivar and state, I must, at the risk of seeming small-spirited, insist on my claim of paternity of this argument, inasmuch as the authors seem unaware of its articulation six years earlier.

24. McNeill, "Frederick the Great."

25. Adams, "An Interdisciplinary Overview of a Mesopotamian City."

26. Lewis, *The Early Chinese Empires*, 6.

27. Heather, *The Fall of the Roman Empire*, 56.

28. Lindner, *Nomads and Ottomans in Medieval Anatolia*, 65.

29. Yoffee and Cowgill, *The Collapse of Ancient States*, 49. Seth Richardson (personal communication) notes that the text for this quotation is a literary piece addressed to the gods and likely to be unrepresentative.

30. Porter, *Mobile Pastoralism*, 324. The term "wall" may be misleading, inasmuch as it may well refer to a string of settlements—fortified or unfortified—marking the limit of political control and conceptualized as a state boundary or perimeter.

31. Wang Haicheng, *Writing and the Ancient State*, 98.

32. There was apparently, prior to state formation, a proto-cuneiform in use a few centuries earlier in large urban institutions—presumably temples—for recording transactions and distributions. David Wengrow, personal communication, May 2015.

33. Nissen, "The Emergence of Writing in the Ancient Near East." Nissen adds, "The emergence of writing as here elaborated, should by no means lead one to proclaim the invention of writing as one of the great intellectual steps taken by mankind. Its impact on intellectual life was not so sudden as to justify the differentiating of a dark 'pre-historic' age from bright history. By the time writing appeared, most of the steps toward a higher, civilized form of living had been taken. Writing appears merely as a by-product along the course of rapid development towards a complex life in towns and states" (360). See also Pollock, *Ancient Mesopotamia*, 168, who also claims that cuneiform was not used for temple hymns, myths, proverbs, and temple dedications until at least 2,500 BCE.

34. Crawford, *Ur*, 88.

35. Algaze, "Initial Social Complexity in Southwestern Asia."

36. This account of early writing in China is drawn largely from Wang Haicheng, *Writing and the Ancient State*, and Lewis, *The Early Chinese Empires*.

37. Lewis, *The Early Chinese Empires*, 274.

38. Algaze, "Initial Social Complexity in Southwestern Asia," 220–222, quoting C. C. Lambert-Karlovsky. See also Scott, *The Art of Not Being Governed*, 220–237.

5. POPULATION CONTROL

1. Steinkeller and Hudson, "Introduction: Labor in the Early States: An Early Mesopotamian Perspective," *Labor in the Ancient World*, 1–35.

2. Sahlins, *Stone Age Economics*.

3. Chayanov, *The Theory of Peasant Economy*, 1–28. Much the same logic is behind the frequently observed "backward bending supply curve for labor" in which precapitalist peoples will engage in wage work with a particular objective (sometimes called a "target income") in mind (wedding expenses, the purchase of a mule) and will, contrary to standard microeconomic logic, work less when the wage is higher, as they will meet their objective that much sooner.

4. Boserup, *The Conditions of Agricultural Growth*, 73.

5. In agrarian societies, the patriarchal family is something of a microcosm of this situation. Holding onto the labor—physical and reproductive—of the women in the family as well as the labor of the children is central to its success, especially the success of its CEO, the patriarch!

6. Thucydides, *The Peloponnesian War*, 221.

7. Richardson, "Early Mesopotamia," 9, 20. The verb "to herd" is, I think, not inadvertent; inasmuch as absconding subjects are compared to "a scattered herd of cattle" (29). Even the wars between the major states had the purpose of reducing the enemy's manpower, the key to successful statecraft (21–22).

8. Santos-Granero, *Vital Enemies*.

9. Hochschild, *Bury the Chains*, 2.

10. For the relationship of state building to slavery and slave raiding, see my *The Art of Not Being Governed*, 85–94.

11. Finley, "Was Greek Civilization Based on Slave Labour?"

12. Ibid., 164.

13. The account immediately below is drawn from Yoffee, *Myths of the Archaic State;* Yoffee and Cowgill, *The Collapse of the Ancient States and Civilizations;* Adams, "An Interdisciplinary Overview of a Mesopotamian City"; Algaze, "Initial Social Complexity in Southwestern Asia"; McCorriston, "The Fiber Revolution."

14. But for a view more in line with my reading, see Diakanoff, *Structure of Society and State in Early Dynastic Sumer.*

15. Gelb, "Prisoners of War in Early Mesopotamia."

16. Tate Paulette examines this process of assessment, collection, and storage in detail, particularly for the third-millennium alluvium settlement Fara, in "Grain, Storage, and State-Making in Mesopotamia."

17. Algaze, "The End of Prehistory and the Uruk Period," 81. Algaze is relying here on R. K. Englund, "Texts from the Late Uruk Period," in Josef Bauer, Robert K. Englund, and Manfred Krebernik, eds., *Mesopotamien: Späturuk-Zeit und frühdynastische Zeit* (Freiburg: Universitätsverlag, 1998), 236.

18. Algaze, "The End of History and the Uruk Period," 81.

19. The conventional Romanization of the cuneiform term is "[e$_2$ asīrī]."

20. Seri, *The House of Prisoners,* 259. The date is two centuries after Ur III, and the circumstances are somewhat exceptional, but I am assuming that many of the practices described bear a family resemblance to earlier practices; the rest of the paragraph is drawn from her account.

21. Nissen and Heine, *From Mesopotamia to Iraq,* 31.

22. Gelb, "Prisoners of War in Early Mesopotamia," 90; and, later but perhaps relevant, Tenney, *Life at the Bottom of Babylonian Society,* 114, 133.

23. Tenney, *Life at the Bottom of Babylonian Society,* 105, 107–118.

24. Piotr Steinkeller, "The Employment of Labor on National Building Projects in the Ur III Period," in Steinkeller and Hudson, *Labor in the Ancient World,* 137–236. Steinkeller and others, it should be added, take a rosy view of major monumental building projects, treating them as festive interludes during which the workforce was well fed and given plenty of entertainment and drink—rather like the cooperative harvest rituals found in the anthropological literature.

25. See, for example, Menu, "Captifs de guerre et dépendance rurale dans l'Égypte du Nouvel Empire"; Lehner, "Labor and the Pyramids"; and Goelet, "Problems of Authority, Compulsion, and Compensation."

26. Quoted in Goelet, "Problems of Authority, Compulsion, and Compensation," 570.

27. Nemet-Rejat, *Daily Life in Ancient Mesopotamia*, 188.

28. The event was during the reign of Ramses III. Quoted in Maria Golia, "After Tahrir," *Times Literary Supplement*, February 12, 2016, p. 14.

29. The account immediately below owes much to Lewis, *The Early Chinese Empires*; Keightley, *The Origins of Chinese Civilization*; and Yates, "Slavery in Early China."

30. See, for example, Yates, "Slavery in Early China."

31. Readers will perhaps have noted that mass migration to northern Europe and North America, though largely voluntary, accomplishes much the same thing in terms of making the productive life of people raised and trained elsewhere available to the country where they settle.

32. Taylor, "Believing the Ancients." For a dissent from this position, see Scheidel, "Quantifying the Sources of Slaves."

33. Rather than a victory, the battle seems actually to have been a standoff, although the term "Armageddon" comes to us from the clash.

34. Thucydides, *The Peloponnesian War*, 173.

35. Cameron, "Captives and Culture Change."

36. See, especially, Steinkeller, "The Employment of Labor on National Building Projects"; Richardson, "Building Larsa"; Dietler and Herbich, "Feasts and Labor Mobilization." Richardson establishes that the amount of labor required to build, say, a city wall was a good deal less than commonly supposed. It is impossible, on the other hand, to determine the quotidian conditions of labor from the self-inflating official declarations of the sumptuous feasts given to "the people" on the completion of a temple. The social bedrock of these arguments rests on the relative ease of flight by discontented subjects. This perspective overlooks the measures taken against flight, as well as the possible ease of capturing replacements by war or purchase.

37. Algaze, "The Uruk Expansion."

38. Oded, *Mass Deportations and Deportees*. On the practice in early Mesopotamia, see Gelb, "Prisoners of War in Early Mesopotamia."

39. Oded, *Mass Deportations and Deportees*, 20. The scribes report

4.5 million deportees over three hundred years, though those figures seem to be grossly inflated by imperial bluster.

40. Nissen and Heine, *From Mesopotamia to Iraq*, 80.

41. Tocqueville, *Democracy in America*, 544; quoted in Darwin, *After Tamerlane*, 24. Tocqueville adds, "Oppression has, at one stroke, deprived the descendants of the Africans of almost all the privileges of humanity." For a similar analogy between animal and human domestication, see also the remarkable book by Reviel Netz, *Barbed Wire*, 15. For a brilliant analysis of the analogy between domesticated animals and slaves in the antebellum U. S. South, see Jacoby, "Slaves by Nature."

6. FRAGILITY OF THE EARLY STATE

1. Adams, "Strategies of Maximization, Stability, and Resilience."

2. Yoffee and Cowgill, *The Collapse of Ancient States and Civilizations*, and McAnany and Yoffee, *Questioning Collapse*.

3. Broodbank, *The Making of the Middle Sea*, 356.

4. For Mycenaean Greece, David Small argues that "collapse" was actually a "devolution" into the smaller and more stable units of small-scale lineages that remained intact and were the building blocks of the larger political formations; "Surviving the Collapse."

5. Yoffee and Cowgill, *The Collapse of Ancient States and Civilizations*, 30, 60.

6. Nissen, *The Early History of the Ancient Near East*, 187.

7. Brinkman, "Settlement Surveys and Documentary Evidence."

8. Algaze, "The Uruk Expansion," and Wengrow, *What Makes Civilization*, 75–82.

9. See Harrison, *Contagion*, for a history of quarantine.

10. Morris, *Why the West Rules—for Now*, 217.

11. Better known as the Antonine plague. Cunliffe, *Europe Between the Oceans*, 393.

12. See in this connection the important work of Radkau, *Nature and Power*; Meiggs, *Trees and Timber in the Ancient Mediterranean World*; and Hughes, *The Mediterranean*.

13. McMahon, "North Mesopotamia in the Third Millennium BC." For a description of the woodland assemblage of the Upper Euphrates, see Moore, Hillman, and Legge, *Village on the Euphrates*, 51–63.

14. Deacon, "Deforestation and Ownership."

15. Mithen, *After the Ice*, 87.

16. See the comparative figures for relative loss of soil and precipitation runoff for "bare soil," "sown with millet," "grassland," and "ungrazed thicket" in Redman, *Human Impact on Ancient Environments*, 101.

17. Mithen, *After the Ice*, 50.

18. McNeill, *Mountains of the Mediterranean World*, 73–75.

19. Artzy and Hillel, "A Defense of the Theory of Progressive Salinization."

20. Adams, "Strategies of Maximization, Stability, and Resilience."

21. Nissen and Heine, *From Mesopotamia to Iraq*, 71.

22. Thucydides, *The Peloponnesian War,* 485. Thucydides also refers to the defection of disillusioned soldiers who had thought they would make money from the campaign without having to fight.

23. The Athenian confederacy was, one might well argue, put in jeopardy by measures of desperation more than a decade earlier. In 425 BCE the Athenians tripled the levies of material and men from their tributaries, this increasing the odds of desertion.

24. I owe this insight to Victor Lieberman; see his *Strange Parallels*, 1: 1–40.

25. A noted metaphor of my ex-colleague Ed Lindblom.

26. Yoffee and Cowgill, *The Collapse of Ancient States and Civilizations*, 260.

27. Quoted in Morris, *Why the West Rules—for Now*, 194.

28. David O'Connor, "Society and Individual in Early Egypt," in Richards and van Buren, *Order, Legitimacy, and Wealth in Ancient States*, 21–35.

29. Ibid., and Broodbank, *The Making of the Middle Sea*, 277.

30. Here I elaborate on the general line of skepticism originally developed in Yoffee and Cowgill, *The Collapse of Ancient States and Civilizations*, and McAnany and Yoffee, *Questioning Collapse*.

31. Tainter, *The Collapse of Complex Societies*.

32. See G. W. Bowersock, "The Dissolution of the Roman Empire," in Yoffee and Cowgill, *The Collapse of Ancient States and Civilizations*, 165–175. Bowersock claims that the Empire disappeared only with the later Arab invasion.

33. Cunliffe, *Europe Between the Oceans*, 364.

34. Riehl, "Variability in Ancient Near Eastern Environmental and Agricultural Development."

35. Adams, "Strategies of Maximization, Stability, and Resilience," 334.

36. Adams, *The Land Behind Bagdad*, 55.

37. Broodbank, *The Making of the Middle Sea*, 349.

38. Richardson, "Early Mesopotamia," 16.

39. "Indeed, the land turns round as does a potter's wheel. The robber possesses riches . . ."; Bell, "The Dark Ages in Ancient History," 75.

40. McNeill, *Plagues and People*, 58–71. David Wengrow (personal communication) believes that the contact via trade and exchange throughout the area would have worked against the isolation of populations that makes possible epidemics among immunologically "naïve" populations. While this is surely true for the major population centers and the trade routes between them, it may be less true for nonstate peoples off the major trade routes and living in populations small enough that many of the common infectious diseases would not have become endemic. McNeill's conjecture remains just that and awaits further investigation.

7. THE GOLDEN AGE OF THE BARBARIANS

1. By "taxation" I mean any more or less regular charge on the production, labor, or revenue of subjects. In early states, "taxes" are likely to take the form of levies in kind (for example, from the harvest of cultivators) or the form of labor (corvée).

2. My colleague Peter Perdue, an expert on the China borderland and nonstate people generally, would put the terminal date later, at the end of the eighteenth century, when, he observes, "nearly all the frontiers of the globe had been occupied by settlers and merchants, and global commodity traders were extracting resources from all the major continents"; personal communication.

3. J. N. Postgate distinguishes, in the Mesopotamian case, "mountain" raids as compared with "pastoralist" raids, terming the latter as more likely to destroy the state; *Early Mesopotamia*, 9.

4. Skaria, *Hybrid Histories*, 132.

5. Cunliffe, *Europe Between the Oceans*, 229.

6. For a useful summary of what we know about the "sea people" and what is in dispute, see Gitin, Mazar, and Stern, *Mediterranean Peoples in Transition*.

7. Cunliffe, *Europe Between the Oceans*, 331.

8. Bronson, "The Role of Barbarians in the Fall of States," 208.

9. Lattimore, "The Frontier in History," 486.

10. Bronson, "The Role of Barbarians in the Fall of States," 200.

11. Porter, *Mobile Pastoralism*, 324. As Porter has also shown, the Amorites were more a branch of Mesopotamian society than "barbarians." They were, to be sure, challengers and usurpers but they were not "outsiders" (61).

12. Burns, *Rome and the Barbarians*, 150.

13. Quoted in volume 1 of Coatsworth et al., *Global Connections*, 76.

14. Clastres, *La Société contre l'État*.

15. Beckwith, *Empires of the Silk Road*, 76.

16. Lattimore, "The Frontier in History," 476–481.

17. Ibid., quoting E. A. Thompson, *A History of Attila and the Huns* (Oxford: Oxford University Press, 1948), 185–186.

18. Lattimore, "The Frontier in History," 481.

19. Herwig Wolfram, *History of the Goths*, trans. Thomas J. Dunlap (Berkeley: University of California Press, 1988), 8, quoted in Beckwith, *Empires of the Silk Road*, 333.

20. Spartacus and his rebels, it should be noted, were seeking to leave Italy but were stopped by treachery and, finally, by Sulla's army. For a history of state-fleeing practices in upland Southeast Asia, see my *The Art of Not Being Governed*.

21. Cunliffe, *Europe Between the Oceans*, 238.

22. Beckwith, *Empires of the Silk Road*, 333–334.

23. Wengrow, *What Makes Civilization*, 99.

24. One could argue, analogously, that large herd animals, by virtue of being relatively "sedentary" and assembling in large numbers at certain times of the year, were uniquely vulnerable to "raiding," aka "hunting," by Homo sapiens with dogs, spears, and bows and hence likely to be among the among the first species to be threatened with extinction as soon as the population of such hunters became numerous.

25. Beckwith, *Empires of the Silk Road*, 321.

26. Santos-Granero, *Vital Enemies.*

27. Perdue reminds me that the relationship between mobile raiders and sedentary creatures may also be found in the animal and insect kingdoms. They are different and, to some degree, competitive subsistence strategies.

28. Owen Lattimore, "On the Wickedness of Being Nomads."

29. Quoted in Beckwith, *Empires of the Silk Road*, 69.

30. Paul Astrom, "Continuity and Discontinuity: Indigenous and Foreign Elements in Cyprus Around 1200 BC," in Gitin, Mazar, and Stern, *Mediterranean Peoples in Transition*, 80–86, quotation on 83.

31. Susan Sherratt, "'Sea Peoples' and the Economic Structure of the Late Second Millennium in the Eastern Mediterranean," in Gitin, Mazar, and Stern, *Mediterranean Peoples in Transition*, 292–313, quotation on 305.

32. This logic is worked out nicely by Charles Tilly in "War Making and State Making as Organized Crime."

33. William Irons, "Cultural Capital, Livestock Raiding."

34. Barfield, "Tribe and State Relations," 169–170.

35. Flannery, "Origins and Ecological Effect of Early Domestication."

36. Broodbank, *The Making of the Middle Sea*, 358. See also the elegant schematic application of this logic to the traditional riverine statelets in the Malay world in Bronson, "Exchange at the Upstream and Downstream Ends."

37. Beckwith, *Empires of the Silk Road*, 328–329. See also Di Cosmo, *Ancient China and Its Enemies.*

38. Fletcher, "The Mongols," 42.

39. Cunliffe, *Europe Between the Oceans*, 378.

40. Ibid., especially chapter 7.

41. Tsing, *The Mushroom at the End of the World.*

42. Beckwith, *Empires of the Silk Road*, 327–328.

43. Artzy, "Routes, Trade, Boats and 'Nomads of the Sea,'" 439–448.

44. Lattimore, "The Frontier in History," 504.

45. Fletcher distinguishes between, on the one hand, "steppe" nomads, who interact far less with settled peoples and agrarian states and for whom raiding is as important as trading, and, on the other, "desert"

nomads, who are more likely to have routine trading relations with sedentary communities and urban society; Fletcher, "The Mongols," 41.

46. Barfield, "The Shadow Empires."

47. See, in this connection, Ratchnevsky, *Genghis Khan*, and Hämäläinen, *Comanche Empire*.

48. Ferguson and Whitehead, "The Violent Edge of Empire," 23.

49. Kradin, "Nomadic Empires in Evolutionary Perspective," 504. See also Barfield, "Tribe and State Relations," for a similar view.

Bibliography

Adams, Robert McC. "Agriculture and Urban Life in Early Southwestern Iran." *Science* 136, no. 3511 (1962): 109–122.

———. *The Land Behind Bagdad: A History of Settlement on the Diyala Plains.* Chicago: University of Chicago Press, 1965.

———. "Anthropological Perspectives on Ancient Trade." *Current Anthropology* 15, no. 3 (1974): 141–160.

———. *Heartland of Cities: Surveys of Ancient Settlements and Land Use on the Central Floodplain of the Euphrates.* Chicago: University of Chicago Press, 1974.

———. "Strategies of Maximization, Stability, and Resilience in Mesopotamian Society, Settlement, and Agriculture." *Proceedings of the American Philosophical Society* 122, no. 5 (1978): 329–335.

———. "The Limits of State Power on the Mesopotamian Plain." *Cuneiform Digital Library Bulletin* 1 (2007).

———. "An Interdisciplinary Overview of a Mesopotamian City and Its Hinterland." *Cuneiform Digital Library Journal* 1 (2008): 1–23.

Algaze, Guillermo. "The Uruk Expansion: Cross Cultural Exchange in Early Mesopotamian Civilization." *Current Anthropology* 30, no. 5 (1989): 571–608.

———. "Initial Social Complexity in Southwestern Asia: The Mesopotamian Advantage." *Current Anthropology* 42, no. 2 (2001): 199–233.

———. "The End of Prehistory and the Uruk Period." In Crawford, *The Sumerian World*, 68–94.

Appuhn, Karl. "Inventing Nature: Forests, Forestry, and State Power in Renaissance Venice." *Journal of Modern History* 72, no. 4 (2000): 861–889.

Armelagos, George J., and Alan McArdle. "Population, Disease, and Evolution." *Memoirs of the Society of American Archaeology*, no. 30 (1975), *Population Studies in Archaeology and Biological Anthropology: A Symposium*, 1–10.

Armelagos, George J., et al. "The Origins of Agriculture: Population Growth During a Period of Declining Health." *Population and Environment: A Journal of Interdisciplinary Studies* 13, no. 1 (1981): 9–22.

Artzy, Michal. "Routes, Trade, Boats and 'Nomads of the Sea.'" In Gitin et al., *Mediterranean Peoples in Transition*, 439–448.

Artzy, Michal, and Daniel Hillel. "A Defense of the Theory of Progressive Salinization in Ancient Southern Mesopotamia." *Geoarchaeology* 3, no. 3 (1988): 235–238.

Asher-Greve, Julia M. "Women and Agency: A Survey from Late Uruk to the End of Ur III." In Crawford, *The Sumerian World*, 345–358.

Asouti, Eleni, and Dorian Q. Fuller. "A Contextual Approach to the Emergence of Agriculture in Southwest Asia: Reconstructing Early Neolithic Plant-food Production." *Current Anthropology* 54, no. 3 (2013): 299–345.

Axtell, James. "The White Indians of Colonial America." *William and Mary Quarterly* 3rd ser. 32 (1975): 55–88.

Bairoch, Paul. *Cities and Economic Development: From the Dawn of History to the Present*. Trans. Christopher Braider. Chicago: University of Chicago Press, 1988.

Baker, Paul T., and William T. Sanders. "Demographic Studies in Anthropology." *Annual Review of Anthropology* 1 (1972): 151–178.

Barfield, Thomas J. "Tribe and State Relations: The Inner Asian Perspective." In Philip S. Khoury and Joseph Kostiner, eds., *Tribes and State Formation in the Middle East*, 153–182. Berkeley: University of California Press, 1990.

———. "The Shadow Empires: Imperial State Formation Along the Chinese Nomad Frontier." In Susan E. Alcock, Terrance N. D'Altroy, et al., eds. *Empires: Perspectives from Archaeology and History*, 11-41. Cambridge: Cambridge University Press, 2001.

Beckwith, Christopher. *Empires of the Silk Road: A History of Central Eurasia from the Bronze Age to the Present*. Princeton: Princeton University Press, 2009.

Bell, Barbara. "The Dark Ages in Ancient History: 1. The First Dark Age in Egypt." *American Journal of Archaeology* 75, no. 1 (1971): 1-26.

Bellwood, Peter. *First Farmers: The Origins of Agricultural Societies*. Oxford: Blackwell, 2005.

Bennet, John. "The Aegean Bronze Age." In Scheidel et al., *Cambridge Economic History*, 175-210.

Berelov, Ilya. "Signs of Sedentism and Mobility in Agro-Pastoral Community During the Levantine Middle Bronze Age: Interpreting Site Function and Occupation Strategy at Zahrat adh-Dhra 1. *Journal of Anthropological Archaeology* 25 (2006): 117-143.

Bernbeck, Reinhard. "Lasting Alliances and Emerging Competition: Economics Developments in Early Mesopotamia." *Journal of Anthropological Archaeology* 14 (1995): 1-25.

Blanton, Richard, and Lane Fargher. *Collective Action in the Formation of Pre-Modern States*. New York: Springer, 2008.

Blinman, Eric. "2000 Years of Cultural Adaptation to Climate Change in the Southwestern United States." *AMBO: A Journal of the Human Environment* 37, sp. 14 (2000): 489-497.

Bocquet-Appel, Jean-Pierre. "Paleoanthropological Traces of a Neolithic Demographic Transition." *Current Anthropology* 43, no. 4 (2002): 637-650.

———. "The Agricultural Demographic Transition (ADT) During and After the Agricultural Inventions." *Current Anthropology* 52, no. S4 (2011): 497-510.

Boone, James L. "Subsistence Strategies and Early Human Population History: An Evolutionary Perspective." *World Archaeology* 34, no. 1 (2002): 6-25.

Boserup, Ester. *The Conditions of Agricultural Growth: The Economics of Agrarian Change Under Population Pressure*. Chicago: Aldine, 1965.

Boyden, S. V. *The Impact of Civilisation on the Biology of Man*. Toronto: University of Toronto Press, 1970.

Braund, D. C., and G. R. Tsetkhladze. "The Export of Slaves from Colchis." *Classical Quarterly* new ser. 39, no. 1 (1988): 114–125.

Brinkman, John Anthony. "Settlement Surveys and Documentary Evidence: Regional Variation and Secular Trends in Mesopotamian Demography." *Journal of Near Eastern Studies* 43, no. 3 (1984): 169–180.

Brody, Hugh. *The Other Side of Eden: Hunters, Farmers, and the Shaping of the World*. Vancouver: Douglas and McIntyre, 2002.

Bronson, Bennett. "Exchange at the Upstream and Downstream Ends: Notes Toward a Functional Model of the Coastal State in Southeast Asia." In Karl Hutterer, ed., *Economic Exchange and Social Interaction in Southeast Asia: Perspectives from Prehistory, History, and Ethnography*, 39–52. Ann Arbor: Center for South and Southeast Asian Studies, University of Michigan, 1977.

———. "The Role of Barbarians in the Fall of States." In Yoffee and Cowgill, *Collapse of Ancient States*, 196–218.

Broodbank, Cyprian. *The Making of the Middle Sea: A History of the Mediterranean from the Beginning to the Emergence of the Classical World*. London: Thames and Hudson, 2013.

Burke, Edmund, and Kenneth Pomeranz, eds. *The Environment and World History*. Berkeley: University of California Press, 2009.

Burnet, Sir MacFarlane, and David O. White. *The Natural History of Infectious Disease*, 4th ed. Cambridge: Cambridge University Press, 1972.

Burns, Thomas S. *Rome and the Barbarians, 100 BC–AD 400*. Baltimore: Johns Hopkins University Press, 2003.

Cameron, Catherine M. "Captives and Culture Change." *Current Anthropology* 52, no. 2 (2011): 169–209.

Cameron, Catherine M., and Steve A. Tomka. *Abandonment of Settlements and Regions: Ethnoarchaeological and Archaeological Approaches*. New Directions in Archaeology. Cambridge: Cambridge University Press, 1996.

Carmichael, G. "Infection, Hidden Hunger. and History." In "Hunger and History: The Impact of Changing Food Production and Consumption Patterns on Society," *Journal of Interdisciplinary History* 14, no. 2 (1983): 249–264.

Carmona, Salvador, and Mahmoud Ezzamel. "Accounting and Forms of

Accountability in Ancient Civilizations: Mesopotamia and Ancient Egypt." Working Paper, Annual Conference of the European Accounting Association, Goteborg, Sweden, 2005.

Carneiro, R. "A Theory of the Origin of the State." *Science* 169 (1970): 733–739.

Chakrabarty, Dipesh. "The Climate of History: Four Theses." *Critical Inquiry* 35 (2009): 197–222.

Chang, Kwang-chih. "Ancient Trade as Economics or as Ecology." In Jeremy Sabloff and C. C. Lamberg-Karlovsky, eds., *Ancient Civilization and Trade*, 211–224. Albuquerque: School of American Research, University of New Mexico Press, 1975.

Chapman, Robert. *Archaeology of Complexity*. London: Routledge, 2003.

Chayanov, A. V. *The Theory of Peasant Economy*. Ed. Daniel Thorner, Basile Kerblay, and R. E. F. Smith. Homewood, Ill.: Richard D. Irwin for the American Economic Association, 1966.

Christensen, Peter. *The Decline of Iranshahr: Irrigation and Environments in the History of the Middle East, 500 BC to AD 1500*. Copenhagen: Museum Tusculanum, 1993.

Christian, David. *Maps of Time: An Introduction to Big History*. Berkeley: University of California Press, 2004.

Clarke, Joanne, ed. *Archaeological Perspectives on the Transmission and Transformation of Culture in the Eastern Mediterranean*. Levant Supplementary Series 2. Oxford: Oxbow, 2005.

Clastres, Pierre. *La Société contre l'État*. Paris: Editions de Minuit, 1974.

Coatsworth, John, Juan Cole, et al. *Global Connections: Politics, Exchange, and Social Life in World History*, vol. 1, *To 1500*. Cambridge: Cambridge University Press, 2015.

Cockburn, I. Aiden. "Infectious Diseases in Ancient Populations." *Current Anthropology* 12, no. 1 (1971): 45–62.

Conklin, Harold C. *Hanunóo Agriculture: A Report on an Integral System of Shifting-Agriculture in the Philippines*. Rome: Food and Agriculture Organization of the United Nations, 1957.

Cowgill, George L. "On Causes and Consequences of Ancient and Modern Population Changes." *American Anthropologist* 77, no. 3 (1975): 505–525.

Crawford, Harriet, ed. *The Sumerian World*. London: Routledge, 2013.

———. *Ur: The City of the Moon God*. London: Bloomsbury, 2015.

Cronon, William. *Changes in the Land: Indians, Colonists, and the Ecology of New England*, rev. ed. New York: Hill and Wang, 2003.

Crossley, Pamela Kyle, Helen Siu, and Donald Sutton, eds., *Empire at the Margins: Culture and Frontier in Early Modern China*. Berkeley: University of California Press, 2006.

Crouch, Barry A. "Booty Capitalism and Capitalism's Booty: Slaves and Slavery in Ancient Rome and the American South." *Slavery and Abolition: A Journal of Slave and Post-Slave Studies* 6, no. 1 (1985): 3–24.

Crumley, Carol L. "The Ecology of Conquest: Contrasting Agropastoral and Agricultural Societies' Adaptation to Climatic Change." In Carol L. Crumley, ed., *Historical Ecology: Cultural Knowledge and Changing Landscapes*, 183–201. School of American Research Advanced Seminar Series. Santa Fe, N.M.: School of American Research Press, 1994.

Cunliffe, Barry. *Europe Between the Oceans: Themes and Variations: 9000 BC–AD 1000*. New Haven: Yale University Press, 2008.

Dalfes, H. Nüzhet, George Kukla, and Harvey Weiss. *Third Millennium BC Climate Change and Old World Collapse*. NATO Advanced Science Institutes Series, Series I, Global Environmental Change 49 (2013).

Dark, Petra, and Henry Gent. "Pests and Diseases of Prehistoric Crops: A Yield 'Honeymoon' for Early Grain Crops in Europe?" *Oxford Journal of Archaeology* 20, no. 1 (2001): 59–78.

Darwin, John. *After Tamerlane: The Rise and Fall of Global Empires, 1400–2000*. London: Penguin, 2007.

Deacon, Robert T. "Deforestation and Ownership: Evidence from Historical Accounts and Contemporary Data." *Land Economics* 75, no. 3 (1999): 341–359.

Diakanoff, M. *Structure of Society and State in Early Dynastic Sumer.* Malibu, Calif.: Monographs of the Ancient Near East, 1, no. 3 (1974).

Diamond, Jared. *Guns, Germs, and Steel: The Fates of Human Societies.* New York: Norton, 1977.

Dickson, D. Bruce. "Circumscription by Anthropogenic Environmental Destruction: An Expansion of Carneiro's (1970) Theory of the Origin of the State." *American Antiquity* 52, no. 4 (1987): 709–716.

Di Cosmo, Nicola. "State Formation and Periodization in Inner Asian History." *Journal of World History* 10, no. 1 (1999): 1–40.

———. *Ancient China and Its Enemies: The Rise of Nomadic Power in East Asian History.* Cambridge: Cambridge University Press, 2011.

Dietler, Michael. "The Iron Age in the Western Mediterranean." In Scheidel et al., *Cambridge Economic History,* 242–276.

Dietler, Michael, and Ingrid Herbich. "Feasts and Labor Mobilization: Dissecting a Fundamental Economic Practice." In M. Dietler and Brian Hayden, eds., *Feasts: Archaeological and Ethnographic Perspectives on Food, Politics, and Power,* 240–264. Washington, D.C.: Smithsonian Institution Press, 2001.

Donaldson, Adam. "Peasant and Slave Rebellions in the Roman Republic." Ph.D. diss., University of Arizona, 2012.

D'Souza, Rohan. *Drowned and Dammed: Colonial Capitalism and Flood Control in Eastern India.* New Delhi: Oxford University Press, 2006.

Dyson-Hudson, Rada, and Eric Alden Smith. "Human Territoriality: An Ecological Reassessment." *American Anthropologist* new ser. 890, no. 1 (1973): 21–41.

Eaton, S. Boyd, and Melvin Konner. "Paleolithic Nutrition." *New England Journal of Medicine* 312, no. 5 (1985): 283–290.

Ebrey, Patricia Buckley. *The Cambridge Illustrated History of China.* Cambridge: Cambridge University Press, 1996.

Elias, Norbert. *The Civilizing Process: Sociogenic and Psychogenic Investigations,* rev. ed. Oxford: Blackwell, 1994.

Ellis, Maria de J. "Taxation in Ancient Mesopotamia: The History of the Term Miksu." *Journal of Cuneiform Studies* 26, no. 4 (1974): 211–250.

Elvin, Mark. *Retreat of the Elephants: An Environmental History of China.* New Haven: Yale University Press, 2004.

Endicott, Kirk. "Introduction: Southeast Asia." In Richard B. Lee and Richard Daly, eds., *The Cambridge Encyclopedia of Hunters and Gatherers,* 275–283. Cambridge: Cambridge University Press, 1999.

Eshed, Vered, et al. "Has the Transition to Agriculture Reshaped the Demographic Structure of Prehistoric Populations? New Evidence from the Levant." *American Journal of Physical Anthropology* 124 (2004): 315–329.

Evans-Pritchard, E. E. *The Nuer: A Description of the Modes of Livelihood and Political Institutions of a Nilotic People.* Oxford: Clarendon, 1940.

Evin, Allowen, et al. "The Long and Winding Road: Identifying Pig Domestication Through Molar Size and Shape." *Journal of Archaeological Science* 40 (2013): 735-742.

Farber, Walter. "Health Care and Epidemics in Antiquity: The Example of Ancient Mesopotamia." Lecture, Oriental Institute, June 26, 2006, CHIASMOS, https://www.youtube.com/watch?v=Yw_4 Cghic_w.

Febvre, Lucien. *A Geographical Introduction to History.* Trans. E. G. Mountford and J. H. Paxton. London: Routledge Kegan Paul, 1923.

Feinman, Gary M., and Joyce Marcus. *Archaic States.* Santa Fe, N.M.: School of American Research, 1998.

Fenner, Frank. "The Effects of Changing Social Organization on the Infectious Diseases of Man." In Boyden, *Impact of Civilisation*, 48-68.

Ferguson, R. Brian, and Neil L. Whitehead. "The Violent Edge of Empire." In R. Brian Ferguson and Neil L. Whitehead, eds., *War in the Tribal Zone: Expanding States and Indigenous Warfare*, 1-30. Santa Fe, N.M.: School of American Research, 1992.

Fiennes, R. N. *Zoonoses and the Origins and Ecology of Human Disease.* London: Academic Press, 1978.

Finley, M. I. "Was Greek Civilization Based on Slave Labour?" *Historia: Zeitschrift fur alte geschichte* 8, no. 2 (1959): 145-164.

Fiskesjo, Magnus. "The Barbarian Borderland and the Chinese Imagination: Travelers in Wa Country." *Inner Asia* 5, no. 1 (2002): 81-99.

Flannery, Kent V. "Origins and Ecological Effect of Early Domestication in Iran and the Middle East." In Ucko and Dimbleby, *Domestication and Exploitation*, 73-100.

Fletcher, Joseph. "The Mongols: Ecological and Social Perspectives." *Harvard Journal of Asiatic Studies* 46, no. 1 (1986): 11-50.

French, E. B., and K. A. Wardle, eds. *Problems in Greek Prehistory: Papers Presented at the Centenary Conference of the British School of Archaeology at Athens.* Manchester: Bristol Classical Press, 1986.

Friedman, Jonathan. "Tribes, States, and Transformations: An Association for Social Anthropology Study." In Maurice Bloch, ed., *Marxist Analyses and Social Anthropology*, 161-200. New York: Wiley, 1975.

Fukuyama, Francis. *The Origins of Political Order: From Prehuman Times to the French Revolution*. New York: Farrar, Straus and Giroux, 2011.

Fuller, Dorian Q., et al. "Cultivation and Domestication Has Multiple Origins: Arguments Against the Core Area Hypothesis for the Origins of Agriculture in the Near East." *World Archaeology* 43, no. 4, special issue, Debates in World Archaeology (2011): 628–652.

Gelb, J. J. "Prisoners of War in Early Mesopotamia." *Journal of Near Eastern Studies* 32, no. 12 (1973): 70–98.

Gibson, McGuire, and Robert D. Briggs. "The Organization of Power: Aspects of Bureaucracy in the Ancient Near East." *Studies in Ancient Oriental Civilization*, no. 46. Chicago: Oriental Institute of the University of Chicago, 1987.

Gilbert, Allan S. "Modern Nomads and Prehistoric Pastoralists: The Limits of Analogy." *Journal of the Ancient Near Eastern Society* 7 (1975): 53–71.

Gilman, A. "The Development of Social Stratification in Bronze Age Europe." *Current Anthropology* 22 (1981): 1–23.

Gitin, Seymour, Amihai Mazar, and Ephraim Stern, eds. *Mediterranean Peoples in Transition: Thirteenth to Early Tenth Centuries BCE*. In Honor of Professor Trude Dothan. Jerusalem: Israel Exploration Society, 1998.

Goelet, Ogden. "Problems of Authority, Compulsion, and Compensation in Ancient Egyptian Labor Practices." In Steinkeller and Hudson, *Labor in the Ancient World*, 523–582.

Goring-Morris, A. Nigel, and Anna Belfer-Cohen. "Neolithization Processes in the Levant: The Outer Envelope." *Current Anthropology* 52, no. S4, The Origins of Agriculture: New Data, New Ideas (2011): S195–S208.

Goudsblom, Johan. *Fire and Civilization*. London: Penguin, 1992.

Graeber, David. *Debt: The First 5,000 Years*. London: Melville House, 2011.

Greger, Michael. "The Human/Animal Interface: Emergence and Resurgence of Zoonotic Infectious Diseases." *Critical Reviews in Microbiology* 33 (2007): 243–299.

Grinin, Leonid E., et al., eds. *The Early State, Its Alternatives and Analogues*. Volgograd: "Uchitel," 2004.

Groenen, Martien A. M., et al. "Analysis of Pig Genome Provides Insight into Porcine Domestication and Evolution." *Nature* 491 (2012): 391–398.

Groube, Les. "The Impact of Diseases upon the Emergence of Agriculture." In D. R. Harris, ed., *The Origins and Spread of Agriculture and Pastoralism in Eurasia*, 101–129. Washington, D.C.: Smithsonian Institution Press, 1996.

Halstead, Paul, and John O'Shea, eds. *Bad Year Economics: Cultural Responses to Risk and Uncertainty*. Cambridge: Cambridge University Press, 1989.

Hämäläinen, Pekka. *Comanche Empire*. New Haven: Yale University Press, 2009.

Harari, Yuval Noah. *Sapiens: A Brief History of Humankind*. London: Harvill Secker, 2011.

Harlan, Jack R. *Crops and Man*, 2nd ed. Madison, Wis.: American Society of Agronomy, Crop Science Society of America, 1992.

Harris, David R. *Settling Down and Breaking Ground: Rethinking the Neolithic Revolution*. Amsterdam: Kroon-Voordrachte 12, 1990.

Harris, David R., and Gordon C. Hillman, eds. *Foraging and Farming: The Evolution of Plant Exploitation*. London: Unwin Hyman, 1989.

Harrison, Mark. *Contagion: How Commerce Has Spread Disease*. New Haven: Yale University Press, 2012.

Headland, T. N., "Revisionism in Ecological Anthropology." *Current Anthropology* 38, no. 4 (1997): 43–66.

Headland, T. N. and L. A. Reid. "Hunter-Gatherers and Their Neighbors from Prehistory to the Present." *Current Anthropology* 30, no. 1 (1989): 43–66.

Heather, Peter. *The Fall of the Roman Empire: A New History of Rome and the Barbarians*. Oxford: Oxford University Press, 2006.

Hendrickson, Elizabeth, and Ingolf Thuesen, eds. *Upon This Foundation: The Ubaid Reconsidered*. Copenhagen: Museum Tusculanum Press, Carsten Niebuhr Institute of Ancient Near Eastern Studies.

Hillman, Gordon. "Traditional Husbandry and Processing of Archaic Cereals in Recent Time: The Operations, Products, and Equipment Which Might Feature in Sumerian Texts." *Bulletin of Sumerian Agriculture* 1 (1984): 114–172.

Hochschild, Adam. *Bury the Chains: Prophets and Rebels in the Fight to Free an Empire's Slaves*. New York: Houghton Mifflin, 2015.

Hodder, Ian. *The Domestication of Europe: Structure and Contingency in Neolithic Societies*. Oxford: Blackwell, 1990.

Hole, Frank. "A Monumental Failure: The Collapse of Susa." In Robin A. Carter and Graham Philip, eds., *Beyond the Ubaid: Transformation and Integration of Late Prehistoric Societies of the Middle East*, 221–226. *Studies in Oriental Civilization*, no. 653. Chicago: Oriental Institute of the University of Chicago, 2010.

Houston, Stephen. *The First Writing: Script Invention as History and Process*. Cambridge: Cambridge University Press, 2004.

Hritz, Carrie, and Jennifer Pournelle. "Feeding History: Deltaic Resilience Inherited Practice and Millennia-scale Sustainability." In H. Thomas Foster II, David John Goldstein, and Lisa M. Paciulli, eds., *The Future in the Past: Historical Ecology Applied to Environmental Issues*, 59–85. Columbia: University of South Carolina Press, 2015.

Hughes, J. Donald. *The Mediterranean: An Environmental History*. Santa Barbara: ABC-CLIO, 2005.

Ingold, T. "Foraging for Data, Camping with Theories: Hunter-Gatherers and Nomadic Pastoralists in Archaeology and Anthropology." *Antiquity* 66 (1992): 790–803.

Irons, William G. "Livestock Raiding Among Pastoralists: An Adaptive Interpretation." In *Papers of the Michigan Academy of Science, Arts, and Letters* 383–414. Ann Arbor: University of Michigan Press, 1965.

———. "Cultural Capital, Livestock Raiding, and the Military Advantage of Traditional Pastoralists." In Grinin et al., *The Early State*, 466–475.

Jacobs, Jane. *The Economy of Cities*. New York: Vintage, 1969.

Jacoby, Karl. "Slaves by Nature? Domestic Animals and Human Slaves." *Slavery and Abolition* 18, no. 1 (1994): 89–98.

Jameson, Michael H. "Agriculture and Slavery in Classical Athens." *Classical Journal* 73, no. 2 (1977): 122–145.

Jones, David S. "Virgin Soils Revisited." *William and Mary Quarterly* 3rd ser. 60, no. 4 (2003): 703–742.

Jones, Martin. *Feast: Why Humans Share Food*. Oxford: Oxford University Press, 2007.

Kealhofer, Lisa. "Changing Perceptions of Risk: The Development of Agro-Ecosystems in Southeast Asia." *American Anthropologist* new ser. 104, no. 1 (2002): 178–194.

Keightley, David N., ed. *The Origins of Chinese Civilization.* Berkeley: University of California Press, 1983.

Kennett, Douglas J., and James P. Kennett. "Early State-Formation in Southern Mesopotamia: Sea Levels, Shorelines, and Climate Change." *Journal of Island and Coastal Archaeology* 1 (2006): 67–99.

Khazanov, Anatoly M. "Nomads of the Eurasian Steppes in Historical Retrospective." In Grinin et al., *The Early State,* 476–499.

Kleinman, Arthur M., et al. "Introduction: Avian and Pandemic Influenza: A Bio-Social Approach." *Journal of Infectious Diseases* 197, supplement 1 (2008): S1–S3.

Kovacs, Maureen Gallery, trans. *The Epic of Gilgamesh.* Stanford: Stanford University Press, 1985.

Kradin, Nikolay N. "Nomadic Empires in Evolutionary Perspective." In Grinin et al., *The Early State,* 501–523.

Larson, Gregor. "Ancient DNA, Pig Domestication, and the Spread of the Neolithic into Europe." *Proceedings of the National Academy of Sciences* 104, no. 39 (2007): 15276–15281.

———. "Patterns of East Asian Pig Domestication, Migration, and Turnover Revealed by Modern and Ancient DNA." *Proceedings of the National Academy of Sciences* 107, no. 17 (2010): 7686–7691.

Larson, Gregor, and Dorian Q. Fuller. "The Evolution of Animal Domestication." *Annual Review of Ecology, Evolution, and Systematics* 45 (2014): 115–136.

Lattimore, Owen. "The Frontier in History" and "On the Wickedness of Being Nomads." In *Studies in Frontier History: Collected Papers, 1928–1958,* 469–491 and 415–426, respectively. London: Oxford University Press, 1962.

Leach, Helen M. "Human Domestication Reconsidered." *Current Anthropology* 44, no. 3 (2003): 349–368.

Lee, Richard B. "Population Growth and the Beginnings of Sedentary Life Among the !Kung Bushmen." In Brian Spooner, ed., *Population Growth: Anthropological Implications,* 301–324. Cambridge: MIT Press, 1972. http://www.popline.org/node/517639.

Lee, Richard B., and Richard Daly. *The Cambridge Encyclopedia of Hunters and Gatherers*. Cambridge: Cambridge University Press, 1999.

Lefebvre, Henri. *The Production of Space*. New York: Wiley-Blackwell, 1992.

Lehner, Mark. "Labor and the Pyramids: The Hiet el-Ghurab 'Workers Town' at Giza." In Steinkeller and Hudson, *Labor in the Ancient World*, 396–522.

Lévi-Strauss, Claude. *La Pensée sauvage*. Paris: Plon, 1962.

Lewis, Mark Edward. *The Early Chinese Empires: Qin and Han*. Cambridge: Belknap Press of Harvard University Press, 2007.

Lieberman, Victor. *Strange Parallels: Southeast Asia in Global Context, c. 800–1830*, vol. 1, *Integration on the Mainland*. Cambridge: Cambridge University Press, 2003; vol. 2, *Mainland Mirrors: Europe, Japan, China, Southeast Asia and the Islands*. Cambridge: Cambridge University Press, 2009.

Lindner, Rudi Paul. *Nomads and Ottomans in Medieval Anatolia*. Indiana University Uralic and Altaic Series 144, Stephen Halkovic, ed. Bloomington: Research Institute for Inner Asian Studies, Indiana University, 1983.

Mann, Charles C. *1491: New Revelations of the Americas Before Columbus*. New York: Knopf, 2005.

Manning, Richard. *Against the Grain: How Agriculture Has Hijacked Civilization*. New York: Northpoint, 2004.

Marston, John M. "Archaeological Markers of Agricultural Risk Management." *Journal of Archaeological Anthropology* 30 (2011): 190–205.

Matthews, Roger. *The Archaeology of Mesopotamia: Theories and Approaches*. Oxford: Routledge, 2003.

Mayshar, Joram, Omer Moav, Zvika Neeman, and Luigi Pascali. "Cereals, Appropriability, and Hierarchy." CEPR Discussion Paper 10742 (2015). www.cepr.org/active/publications/discussion_papers/dp.php?dpno=10742.

McAnany, Patricia, and Norman Yoffee, eds. *Questioning Collapse: Human Resilience, Ecological Vulnerability, and the Aftermath of Empire*. Cambridge: Cambridge University Press, 2009.

McCorriston, Joy. "The Fiber Revolution: Textile Extensification, Alienation, and Social Stratification in Ancient Mesopotamia." *Current Anthropology* 38, no. 4 (1997): 517–535.

McKeown, Thomas. *The Origins of Human Disease*. Oxford: Blackwell, 1988.

McLean, Rose B. "Cultural Exchange in Roman Society: Freed Slaves and Social Value." Ph.D. thesis, Princeton University, 2012.

McMahon, Augusta. "North Mesopotamia in the Third Millennium BC." In Crawford, *The Sumerian World*, 462–475.

McNeill, J. R. *Mountains of the Mediterranean World: An Environmental History*. Cambridge: Cambridge University Press, 1992.

———. "The Anthropocene Debates: What, When, Who, and Why?" Paper Presented to the Program in Agrarian Studies Colloquium, Yale University, September 11, 2015.

McNeill, W. H. *Plagues and People*. New York: Monticello Editions, History Book Club, 1976.

———. *The Human Condition: An Ideological and Historical View*. Princeton: Princeton University Press, 1980.

———. "Frederick the Great and the Propagation of Potatoes." In Byron Hollinshead and Theodore K. Rabb, eds., *I Wish I'd Have Been There: Twenty Historians Revisit Key Moments in History*, 176–189. New York: Vintage, 2007.

Meek, R. *Social Science and the Ignoble Savage*. Cambridge: Cambridge University Press, 1976.

Meiggs, Russell. *Trees and Timber in the Ancient Mediterranean World*. Oxford: Oxford University Press, 1982.

Menu, Bernadette. "Captifs de guerre et dépendance rurale dans l'Égypte du Nouvel Empire." In Bernadette Menu, ed., *La Dépendance rurale dans l'Antiquité égyptienne et proche-orientale*. Cairo: Institut Français d'archéologie orientale, 2004.

Mitchell, Peter. *Horse Nations: The Worldwide Impact of the Horse on Indigenous Societies Post 1492*. Oxford: Oxford University Press, 2015.

Mithen, Steven. *After the Ice: A Global Human History, 20,000–5000 BC*. Cambridge: Harvard University Press, 2003.

Moore, A. M. T., G. C. Hillman, and A. J. Legge. *Village on the Euphrates*. Oxford: Oxford University Press, 2000.

Morris, Ian. "Early Iron Age Greece." In Scheidel et al., *Cambridge Economic History*, 211–241.

———. *Why the West Rules—for Now: The Patterns of History and What*

They Reveal About the Future. New York: Farrar, Straus and Giroux, 2010.

Mumford, Jeremy Ravi. *Vertical Empire: The General Resettlement of the Andes*. Durham, N.C.: Duke University Press, 2012.

Nemet-Rejat, Karen Rhea. *Daily Life in Ancient Mesopotamia*. Peabody, Mass.: Hendrickson, 2002.

Netz, Reviel. *Barbed Wire: An Ecology of Modernity*. Middletown, Conn.: Wesleyan University Press, 2004.

Nissen, Hans J. "The Emergence of Writing in the Ancient Near East." *Interdisciplinary Science Reviews* 10, no. 4 (1985): 349–361.

———. *The Early History of the Ancient Near East, 9000–2000 BC*. Chicago: University of Chicago Press, 1988.

Nissen, Hans J., Peter Damerow, and Robert S. Englund. *Ancient Bookkeeping: Early Writing and Techniques of Administration in the Ancient Near East*. Chicago: University of Chicago Press, 1993.

Nissen, Hans J., and Peter Heine. *From Mesopotamia to Iraq: A Concise History*. Trans. Hans J. Nissen. Chicago: University of Chicago Press, 2009.

O'Connor, Richard A. "Agricultural Change and Ethnic Succession in Southeast Asian States: A Case for Regional Anthropology." *Journal of Asian Studies* 54, no. 4 (1995): 968–996.

Oded, Bustenay. *Mass Deportations and Deportees in the Neo-Assyrian Empire*. Weisbaden: Reichert, 1979.

Ottoni, Claudio, et al. "Pig Domestication and Human-Mediated Dispersal in Western Eurasia Revealed Through Ancient DNA and Geometric Morphometrics." *Molecular Biology and Evolution* 30, no. 4 (2012): 824–832.

Padgug, Robert A. "Problems in the Theory of Slavery and Slave Society." *Science and Society* 49, no. 1 (1976): 3–27.

Panter-Brick, Catherina, Robert H. Layton, and Peter Rowley-Conwy, eds. *Hunter-Gatherers: An Interdisciplinary Perspective*. Cambridge: Cambridge University Press, 2001.

Park, Thomas. "Early Trends Toward Class Stratification: Chaos, Common Property, and Flood Recession Agriculture." *American Anthropologist* 94 (1992): 90–117.

Paulette, Tate. "Grain, Storage, and State-Making in Mesopotamia, 3200–2000 BC." In Linda R. Manzanilla and Mitchel S. Rothman, eds., *Storage in Complex Societies: Administration, Organization, and Control*, 85–109. London: Routledge, 2016.

Perdue, Peter C. *Exhausting the Earth: State and Peasant in Hunan, 1500–1850 AD*. Cambridge: Harvard University Press, 1987.

———. *China Marches West: The Ching Conquest of Central Eurasia*. Cambridge: Harvard University Press, 2005.

Pinker, Steven. *The Better Angels of Our Nature: Why Violence Has Declined*. New York: Penguin, 2011.

Pollan, Michael. *The Botany of Desire: A Plant's-Eye View of the World*. New York: Random House, 2001.

Pollock, Susan. "Bureaucrats and Managers, Peasants and Pastoralists, Imperialists and Traders: Research on the Uruk and Jemdet Nasr Periods in Mesopotamia." *Journal of World Prehistory* 6, no. 3 (1992): 297–336.

———. *Ancient Mesopotamia: The Eden That Never Was*. Cambridge: Cambridge University Press, 1999.

Ponting, Clive. *A Green History of the World: The Environment and the Collapse of Great Civilizations*. New York: Penguin, 1993.

Porter, Anne. *Mobile Pastoralism and the Formation of Near Eastern Civilization: Weaving Together Societies*. Cambridge: Cambridge University Press, 2012.

Possehl, Gregory L. "The Mohenjo-Daro Floods: A Reply." *American Anthropologist* 69, no. 1 (1967): 32–40.

Postgate, J. N. *Early Mesopotamia: Society and Economy at the Dawn of History*. London: Routledge, 1992.

———. "A Sumerian City: Town and Country in the 3rd Millennium B.C." *Scienza dell'Antichita Storia Archaeologia* 6–7 (1996): 409–435.

Pournelle, Jennifer. "Marshland of Cities: Deltaic Landscapes and the Evolution of Early Mesopotamian Civilization." Ph.D. thesis, University of California at San Diego, 2003.

———. "Physical Geography." In Crawford, *The Sumerian World*, 13–32.

Pournelle, Jennifer, and Guillermo Algaze. "Travels in Edin: Deltaic Resilience and Early Urbanism in Greater Mesopotamia." In H. Crawford et al., eds., *Preludes to Urbanism: Studies in the Late Chalcolithic of Mesopotamia in Honour of Joan Oates*, 7–34. Oxford: Archaeopress, 2010.

Pournelle, Jennifer, Nagham Darweesh, and Carrie Hritz. "Resilient Landscapes: Riparian Evolution in the Wetlands of Southern Iraq." In Dan Lawrence, Mark Altaweel, and Graham Philip, eds., *New Agendas in Remote Sensing and Landscape Archaeology in the Near East*. Chicago: Oriental Institute of the University of Chicago, forthcoming.

Price, Richard. *Maroon Societies: Rebel Slave Communities in the Americas*, 2nd ed. Baltimore: Johns Hopkins University Press, 1979.

Pyne, Stephen. *World Fire: The Culture of Fire on Earth*. Seattle: University of Washington Press, 1977.

Radkau, Joachim. *Nature and Power: A Global History of the Environment*. Cambridge: Cambridge University Press, 2008.

Radner, Karen. "Fressen und gefressen werden: Heuschrecken als Katastrophe und Delikatesse im altern Vorderen Orient." *Welt des Orients* 34 (2004): 7–22.

Ratchnevsky, Paul. *Genghis Khan: His Life and Legacy*. Trans. T. N. Haining. London: Wiley-Blackwell, 1993.

Redman, Charles. *Human Impact on Ancient Environments*. Tucson: University of Arizona Press, 1999.

Reid, Anthony. *Southeast Asia in the Age of Commerce*, vol. 1, *The Lands Below the Winds*. New Haven: Yale University Press, 1988.

Renfrew, Colin, and John F. Cherry, eds. *Peer Polity Interaction and Socio-Political Change*. New Directions in Archaeology. Cambridge: Cambridge University Press, 1986.

Richards, Janet, and Mary van Buren. *Order, Legitimacy, and Wealth in Ancient States*. Cambridge: Cambridge University Press, 2000.

Richardson, Seth, ed. *Rebellions and Peripheries in the Cuneiform World*. American Oriental Series 91. New Haven: American Oriental Society, 2010.

———. "Early Mesopotamia: The Presumptive State." *Past and Present*, no. 215 (2012): 3–48.

———. "Building Larsa: Labor-Value, Scale, and Scope-of-Economy in Ancient Mesopotamia." In Steinkeller and Hudson, *Labor in the Ancient World*, 237–328.

Riehl, S. "Variability in Ancient Near Eastern Environmental and Agricultural Development." *Journal of Arid Environments* 86 (2011): 1–9.

Rigg, Jonathan. *The Gift of Water: Water Management, Cosmology, and the State in Southeast Asia*. London: School of Oriental and African Studies, 1992.

Rindos, David. *The Origins of Agriculture: An Evolutionary Perspective*. San Diego: Academic Press, 1984.

Roosevelt, Anna Curtenius. "Population, Health, and the Evolution of Subsistence: Conclusions from the Conference." In M. N. Cohen and G. J. Armelagos, eds., *Paleopathology and the Origins of Agriculture*, 259–283. Orlando: Academic Press, 1984.

Rose, Jeffrey I. "New Light on Human Prehistory in the Arabo-Persian Gulf Oasis." *Current Anthropology* 51, no. 6 (2010): 849–883.

Roth, Eric A. "A Note on the Demographic Concomitants of Sedentism." *American Anthropologist* 87, no. 2 (1985): 380–382.

Rowe, J. H., and John V. Murra. "An Interview with John V. Murra." *Hispanic American Historical Review* 64, no. 4 (1984): 633–653.

Rowley-Conwy, Peter, and Mark Zvelibil. "Saving It for Later: Storage by Prehistoric Hunter-Gatherers in Europe." In Halstead and O'Shea, *Bad Year Economics*, 40–56.

Runnels, Curtis, et al. "Warfare in Neolithic Thessaly: A Case Study." *Hesperia* 78 (2009): 165–194.

Sahlins, Marshall. *Stone Age Economics*. Chicago: Aldine, 1974.

Saller, Richard P. "Household and Gender." In Scheidel et al., *Cambridge Economic History*, 87–112.

Sallers, Robert. "Ecology." In Scheidel et al., *Cambridge Economic History*, 15–37.

Santos-Granero, Fernando. *Vital Enemies: Slavery, Predation, and the Amerindian Political-Economy of Life*. Austin: University of Texas Press, 2009.

Sawyer, Peter. "The Viking Perspective." *Journal of Baltic Studies* 13, no. 3 (1982): 177–184.

Scheidel, Walter. "Quantifying the Sources of Slaves in the Early Roman Empire." *Journal of Roman Studies* 87, no. 19 (1997): 156–169.

———. "Demography." In Scheidel et al., *Cambridge Economic History*, 38–86.

Scheidel, Walter, Ian Morris, and Richard Saller, eds. *The Cambridge Economic History of the Greco-Roman World*. Cambridge: Cambridge University Press, 2007.

Schwartz, Glenn M., and John J. Nichols, eds. *After Collapse: The Regeneration of Complex Societies.* Tucson: University of Arizona Press, 2006.

Scott, James C. *The Art of Not Being Governed: An Anarchist History of Upland Southeast Asia.* New Haven: Yale University Press, 2009.

Seri, Andrea. *The House of Prisoners: Slaves and State in Uruk During the Revolt Against Samsu-iluna.* Boston: de Gruyter, 2013.

Sherratt, Andrew. "Reviving the Grand Narrative: Archaeology and Long-term Change," *Journal of European Archaeology* (1995): 1–32.

———. *Economy and Society in Prehistoric Europe: Changing Perspectives.* Edinburgh: Edinburgh University Press, 1997.

———. "The Origins of Farming in South-West Asia." Archatlas 4.1 (2005), http://www.archatlas.dept.shef.ac.uk/OriginsFarming/Farming.php.

Shipman, Pat. *The Invaders: How Humans and Their Dogs Drove Neanderthals to Extinction.* Cambridge: Belknap Press of Harvard University Press, 2015.

Skaria, Ajay. *Hybrid Histories: Forests, Frontiers, and Wildness in Western India.* Oxford: Oxford University Press, 1999.

Skrynnikova, Tatanya D. "Mongolian Nomadic Society of the Empire Period." In Grinin et al., *The Early State,* 525–535.

Small, David. "Surviving the Collapse: The Oikos and Structural Continuity Between Late Bronze Age and Later Greece." In Gitin et al., *Mediterranean Peoples in Transition,* 283–291.

Smith, Adam T. "Barbarians, Backwaters, and the Civilization Machine: Integration and Interruption Across Asia's Early Bronze Age Landscapes." Keynote Presentation at Asian Dynamics Conference, University of Copenhagen, October 22–24, 2014.

Smith, Bruce D. *The Emergence of Agriculture.* New York: Scientific American Library, 1995.

———. "Low Level Food Production." *Journal of Archaeological Research* 9, no. 1 (2001): 1–43.

Smith, Monica L. "How Ancient Agriculturalists Managed Yield Fluctuations Through Crop Selection and Reliance on Wild Plants: An Example from Central India." *Economic Botany* 60, no. 1 (2006): 39–48.

Starr, Harry. "Subsistence Models and Metaphors for the Transition to

Agriculture in Northwestern Europe." *Michigan Discussions in Anthropology* 15, no. 1 (2005).

Steinkeller, Piotr, and Michael Hudson, eds. *Labor in the Ancient World*, vol. 5, International Scholars Conference on Ancient Near Eastern Economies. Dresden: LISLET Verlag, 2015.

Tainter, Joseph A. *The Collapse of Complex Societies*. Cambridge: Cambridge University Press, 1988.

———. "Archaeology of Overshoot and Collapse." *Annual Review of Anthropology* 35 (2006): 59–74.

Taylor, Timothy. "Believing the Ancients: Quantitative and Qualitative Dimensions of Slavery and the Slave Trade in Later Premodern Eurasia." *World Archaeology* 33, no. 1 (2001): 27–43.

Tenney, Jonathan S. *Life at the Bottom of Babylonian Society: Servile Laborers at Nippur in the 14th and 13th Centuries BC*. Leiden: Brill, 2011.

Thucydides. *The Peloponnesian War*. Trans. Rex Warner. New York: Penguin, 1972.

Tilly, Charles. "War Making and State Making as Organized Crime." In Peter Evans, Dietrich Rueschmeyer, and Theda Skocpol, eds., *Bringing the State Back In*, 169–191. Cambridge: Cambridge University Press, 1985.

Tocqueville, Alexis de. *Democracy in America*, vol. 2. New York: Vintage, 1945.

Trigger, Bruce G. *Understanding Early Civilizations: A Comparative Study*. Cambridge: Cambridge University Press, 2003.

Trut, Lyudmilla. "Early Canine Domestication: The Farm Fox Experiments." *Scientific American* 87, no. 2 (1999): 160–169.

Tsing, Anna Lowenhaupt. *The Mushroom at the End of the World: On the Possibility of Life in Capitalist Ruins*. Princeton: Princeton University Press, 2015.

Ucko, Peter J., and G. W. Dimbleby, eds. *The Domestication and Exploitation of Plants and Animals*. Proceedings of a Meeting of the Research Seminar in Archaeology and Related Subjects held at the Institute of Archaeology, London University. Chicago: Aldine, 1969.

Vansina, Jan. *How Societies Are Born: Governance in West Central Africa before 1600*. Charlottesville: University of Virginia Press, 2004.

Walker, Phillip L. "The Causes of Porotic Hyperostosis and Cribra

Orbitalia: A Reappraisal of the Iron-Deficiency-Anemia Hypothesis." *American Journal of Physical Anthropology* 139 (2009): 109-125.

Wang Haicheng. *Writing and the Ancient State: Early China in Comparative Perspective.* Cambridge: Cambridge University Press, 2014.

Weber, David. *Barbaros: Spaniards and Their Savages in the Age of Enlightenment.* New Haven: Yale University Press, 2005.

Weiss, H., et. al. "The Genesis and Collapse of Third Millennium North Mesopotamian Civilization," *Science* 261 (1993): 995-1004.

Wengrow, David. *The Archaeology of Early Egypt: Social Transformation in North-East Africa, 10,000 to 2,650 BC.* Cambridge: Cambridge University Press, 2006.

———. *What Makes Civilization: The Ancient Near East and the Future of the West.* Oxford: Oxford University Press, 2010.

Wilkinson, Toby C., Susan Sherratt, and John Bennet, eds. *Interweaving Worlds: Systemic Interactions in Eurasia, 7th to 1st Millennia BC.* Oxford: Oxbow, 2011.

Wilkinson, Tony J. "Hydraulic Landscapes and Irrigation Systems of Sumer." In Crawford, *The Sumerian World,* 33-54.

Wilson, Peter J. *The Domestication of the Human Species.* New Haven: Yale University Press, 1988.

Woods, Christopher. *Visible Writing: The Invention of Writing in the Ancient Middle-East and Beyond.* Chicago: University of Chicago Press, 2010.

Wrangham, Richard. *Catching Fire: How Cooking Made Us Human.* New York: Basic, 2009.

Yates, Robin D. S. "Slavery in Early China: A Socio-Cultural Approach." *Journal of East Asian Archaeology* 5, nos. 1-2 (2001): 283-331.

Yoffee, Norman. *Myths of the Archaic State: Evolution of the Earliest Cities, States, and Civilizations.* Cambridge: Cambridge University Press, 2005.

Yoffee, Norman, and George L. Cowgill, eds. *The Collapse of Ancient States and Civilizations.* Tucson: University of Arizona Press, 1988.

Yoffee, Norman, and Brad Crowell, eds., *Excavating Asian History: Interdisciplinary Studies in History and Archaeology.* Tucson: University of Arizona Press, 2006.

Yoffee, Norman, and Andrew Sherratt, eds. *Archaeological Theory: Who Sets the Agenda.* Cambridge: Cambridge University Press, 1993.

Zeder, Melinda A. *Feeding Cities' Specialized Animal Economy in the Ancient Middle East.* Washington, D.C.: Smithsonian Institution Press, 1991.

———. "After the Revolution: Post Neolithic Subsistence in Northern Mesopotamia." *American Anthropologist* new ser. 96, no. 1 (1994): 97-126.

———. "The Origins of Agriculture in the Near East." *Current Anthropology* 52, no. S4 (2011): S221-S235.

———. "The Broad Spectrum Revolution at 40: Resource Diversity, Intensification, and an Alternative to Optimum Foraging Explanations." *Journal of Anthropological Archaeology* 321 (2012): 241-264.

———. "Pathways to Animal Domestication." In P. Gepts, T. R. Famula, R. L. Bettinger, et al., eds., *Biodiversity in Agriculture: Domestication, Evolution, and Sustainability*, 227-259. Cambridge: Cambridge University Press, 2012.

Zeder, Melinda A., Eve Emshwiller, Bruce D. Smith, and Daniel Bradley. "Documenting Domestication: The Intersection of Genetics and Archaeology." *Trends in Genetics* 22, no. 3 (2016): 139-155.

Index

Page numbers in *italics* refer to illustrations.